Idealization and the Aims of Science

Idealization and the Aims of Science

ANGELA POTOCHNIK

The University of Chicago Press
Chicago and London

The University of Chicago Press, Chicago 60637
The University of Chicago Press, Ltd., London
© 2017 by The University of Chicago
All rights reserved. No part of this book may be used or reproduced in any manner whatsoever without written permission, except in the case of brief quotations in critical articles and reviews. For more information, contact the University of Chicago Press, 1427 E. 60th St., Chicago, IL 60637.
Published 2017
Paperback edition 2020

29 28 27 26 25 24 23 22 21 20   1 2 3 4 5

ISBN-13: 978-0-226-50705-7 (cloth)
ISBN-13: 978-0-226-75944-9 (paper)
ISBN-13: 978-0-226-50719-4 (e-book)
DOI: https://doi.org/10.7208/chicago/9780226507194.001.0001

Library of Congress Cataloging-in-Publication Data

Names: Potochnik, Angela, author.
Title: Idealization and the aims of science / Angela Potochnik.
Description: Chicago : The University of Chicago Press, 2017. | Includes bibliographical references and index.
Identifiers: LCCN 2017028476 | ISBN 9780226507057 (cloth : alk. paper) | ISBN 9780226507194 (e-book)
Subjects: LCSH: Science—Philosophy. | Idealism.
Classification: LCC Q175.P88155 2017 | DDC 501—dc23
LC record available at https://lccn.loc.gov/2017028476

*For Mabel and Amelia,
the bookends of this book*

Contents

Preface ix

1 Introduction: Doing Science in a Complex World ... 1
   1.1 *Science by Humans* ... 2
   1.2 *Science in a Complex World* ... 11
   1.3 *The Payoff: Idealizations and Many Aims* ... 18

2 Complex Causality and Simplified Representation ... 23
   2.1 *Causal Patterns in the Face of Complexity* ... 24
      2.1.1 *Causal Patterns* ... 24
      2.1.2 *Causal Complexity* ... 35
   2.2 *Simplification by Idealization* ... 41
      2.2.1 *Reasons to Idealize* ... 42
      2.2.2 *Idealizations' Representational Role* ... 50
      2.2.3 *Rampant and Unchecked Idealization* ... 57

3 The Diversity of Scientific Projects ... 62
   3.1 *Broad Patterns: Modeling Cooperation* ... 63
   3.2 *A Specific Phenomenon: Variation in Human Aggression* ... 70
   3.3 *Predictions and Idealizations in the Physical Sciences* ... 80
   3.4 *Surveying the Diversity* ... 88

4 Science Isn't after the Truth ... 90
   4.1 *The Aims of Science* ... 91
      4.1.1 *Understanding as Science's Epistemic Aim* ... 93
      4.1.2 *Separate Pursuit of Science's Aims* ... 104
   4.2 *Understanding, Truth, and Knowledge* ... 112
      4.2.1 *The Nature of Scientific Understanding* ... 112
      4.2.2 *The Role of Truth and Scientific Knowledge* ... 117

5 Causal Pattern Explanations ... 122
   5.1 *Explanation, Communication, and Understanding* ... 123
   5.2 *An Account of Scientific Explanation* ... 134
      5.2.1 *The Scope of Causal Patterns* ... 135

		5.2.2 The Crucial Role of the Audience	145
		5.2.3 Adequate Explanations	153
6	**Levels and Fields of Science**		161
	6.1	Levels in Philosophy and Science	162
	6.2	Going without Levels	170
		6.2.1 Against Hierarchy	170
		6.2.2 Prizing Apart Forms of Stratification	176
	6.3	The Fields of Science and How They Relate	185
7	**Scientific Pluralism and Its Limits**		198
	7.1	The Entrenchment of Social Values	199
	7.2	How Science Doesn't Inform Metaphysics	206
	7.3	Scientific Progress	213

Acknowledgments 223
List of Figures 225
List of Tables 227
Notes 229
References 235
Index 247

Preface

Physicists sometimes assume that surfaces are frictionless planes. Biologists sometimes assume that populations of organisms are infinite in size. And economists sometimes assume that humans are perfectly rational agents. None of these things is true; they are all idealizations. Idealizations are assumptions made without regard for whether they are true and often with full knowledge that they are false. Idealizations of all kinds pervade science, and it's uncommon for scientists to try to replace them with more accurate assumptions. On the face of it, this is a puzzle. Why do scientists deliberately maintain falsehoods in their theories and models? What do idealizations contribute to science?

In this book, I motivate a strong view of idealizations' centrality to science, and I reconsider the aims of science in light of that centrality. On the account I develop, science does not pursue truth directly but instead aims to support human cognitive and practical ends. Those are projects to which idealizations can directly contribute in a number of ways.

The first three chapters are used to develop my account of idealization's central role in science. In Chapter 1, I discuss how science is shaped by its human practitioners and by the world's complexity. Together, these two ideas inspire a view of science as the search for causal patterns, a search that invariably relies heavily on idealizations. Idealizations contribute to science in a variety of ways, including by playing a positive representational role. These ideas are developed in Chapter 2. In Chapter 3, I detail a few case studies that demonstrate the ubiquity of idealization in science, as well as the wide range of purposes it serves.

The last four chapters explore the implications of this account of idealization for central philosophical debates about the aims of science. Chapter 4 motivates the idea that the epistemic aim of science is not truth but human understanding. Understanding is a cognitive achievement, and unlike truth, it can be directly furthered by idealizations. In Chapter 5, I develop an account of scientific explanation that does justice to how the production of understanding depends on human cognizers. Then, in Chapter 6, I challenge classic conceptions of scientific levels of organization, and I develop an opposed view that better accords with idealized representation across all fields of science. Finally, Chapter 7 shows how my account of idealization and the aims of science expands the influence of human characteristics and values on science's aims and products while also constraining scientific and metaphysical pluralism.

This is ultimately a book about how science is influenced by its creators, limited human beings finding our way in a complex world. The science we create liberally employs falsehoods. As a result, scientific knowledge does not in general consist of truths about the phenomena in our world but is more partial and indirect in its relationship to the world, mediated by human concerns. The fields of science do not reflect compositional relationships found in nature but are instead a haphazard division of labor according to researchers' interests. Any successful account of scientific explanation must be shaped by the cognitive requirements for human understanding. And, finally, human aims and values permeate scientific methods and products. Yet none of the views I develop in this book is intended to undermine the scientific enterprise. Quite to the contrary, on my interpretation of what science is designed to accomplish, the success of science becomes all the more obvious.

This book is primarily written for philosophers of science and students of philosophy of science. But I hope it is also of interest to theoretically minded scientists and anyone else who thinks hard about the nature of science. I have tried to address the topics in ways that do not assume familiarity with the specific philosophical debates to which they relate. The ideas put forward in this book have significance for classic issues in philosophy of science, but I believe they also have significance beyond that, including practical implications for scientific practice.

1
Introduction: Doing Science in a Complex World

All scientific research has been accomplished by human agents, for human ends. Each scientist is an individual human being, with a unique combination of characteristics, concerns, and background. The science these individuals pursue is aimed to provide greater human understanding of the world we inhabit and of ourselves and to further human projects of construction, manipulation, and control. Increasingly, philosophers of science have focused on accounting for scientific practice as it actually occurs: the often-messy and never-completed project of limited human beings, pursued for human ends. This stands in contrast to philosophical approaches to science popular in the past that attempted to transcend the messiness of science and the limitations of its current practitioners. One of those approaches is "rationally reconstructing" science to demonstrate its logical or epistemic basis, setting aside historical contingencies and any other deviations from the logically and epistemically ideal. Rudolf Carnap (1928) famously employed this approach in *Der logische Aufbau der Welt*. A second philosophical approach that attempts to transcend science's imperfections is, instead of accounting for today's actual science, aiming to predict what a future, more perfect science will look like. This strategy was explicitly employed by Paul Oppenheim and Hilary Putnam (1958), who developed a "working hypothesis" that all of science will ultimately be grounded explicitly in microphysics.[1]

Along with philosophers' increasing focus on actual scientific practice, there is also greater attention paid to how complexity influences science. Quite many scientists and philosophers of science have been impressed by the complexity of the world we inhabit and investigate, and appreciation for this complexity increasingly shapes scientific approaches as well as philosophical accounts of science. Phenomena occur in seemingly endless variety and permutations.[2] Simple accounts of these phenomena generally meet with only limited success, and approaches to studying complex systems have proliferated across many fields of science. Philosophers have variously called this world disordered, complex, dappled, and unsimple (Dupré 1993; Bechtel and Richardson 1993; Cartwright 1999; Wimsatt 2007; Mitchell 2012b). These philosophers and others have attempted to articulate the implications of this complexity for our theories of science and metaphysics, with wide-ranging results. Our best scientific laws may, strictly speaking, be false (Cartwright 1983). Our categories, including scientific categories, may not "carve nature at its joints" (Dupré 1993). Phenomena may not actually be law governed or predictable (Cartwright 1999). Scientific practices may vary widely, consisting mainly of fallible, heuristic procedures (Wimsatt 2007). The influence of different causes on a phenomenon may not be separable, even in principle (Mitchell 2012b).

These two observations—that science is ultimately the project of limited human beings and that the world we inhabit is incredibly complex—together constitute the starting point of my investigation in this book. Both are familiar ideas in today's philosophy of science. Nonetheless, tracing out their full implications leads to surprising conclusions, conclusions that conflict with a variety of widely held philosophical ideas about science. Most basically, a science practiced by limited human beings in a complex world results in widespread idealization. Idealizations are assumptions made without regard for whether they are true, generally with full knowledge that they are false. Classic examples are the assumption of a frictionless plane in physics and the assumption of perfectly rational agents in economics. Despite their falsity, idealizations appear in most every scientific project and product, for a range of purposes, and they are not eliminated or even controlled for in the ways we might expect. This widespread idealization outstrips what most philosophers are willing to accept. Accordingly, the full scope of the use of idealizations in science has wide-ranging implications for our best theories of science. In this book, I investigate the implications of idealization for what science shows us about the world, levels of organization and the relationship among fields of science, the nature of scientific explanations, the role of human values in science, and even the very aims of science. In this chapter, I begin by considering in greater depth these observations of science as a human project and of the complexity of the world.

1.1 Science by Humans

In *The Descent of Man*, Charles Darwin (1871) observed of animals that "the males are almost always the wooers" and that "the female, on the other hand,

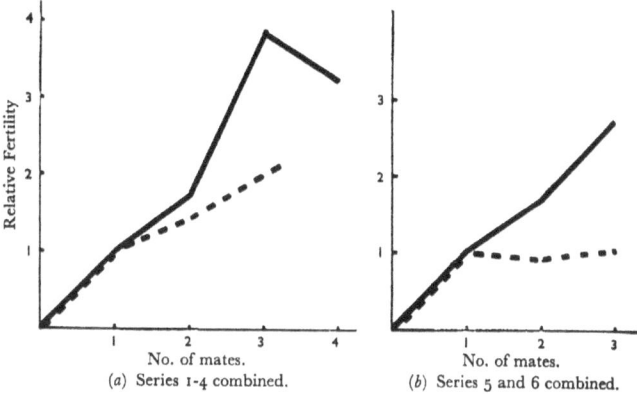

FIGURE 1.1. These are Bateman's original figures, depicting the findings of his experiments for the relationship between the number of matings and reproductive success. The lines compare the data for male fruit flies (solid lines) and female fruit flies (dashed lines). Figures (a) and (b) result from two different series of experiments (Bateman 1948, Fig. 1). Reprinted by permission from Macmillan Publishers Ltd: *Heredity*, copyright 1948.

with the rarest exceptions, is less eager than the male ... she is coy." This idea, that in most species males are aggressive and sexually promiscuous while females are passive and sexually selective, has been widely held by biologists, from Darwin up to today. It is one of the primary ideas of what is called sexual selection theory, one part of evolutionary theory. Male elephant seals fight each other for dominance over harems of females. Peacocks display their colorful trains in the hopes that a peahen will choose to mate with them instead of other peacocks.

A. J. Bateman (1948) provided an explanation for this phenomenon, based on his research on *Drosophila* (fruit flies). According to what has been dubbed Bateman's principle, males stand to produce more offspring as a result of more matings, whereas in most species, females gain few if any additional offspring from multiple matings. If this is true, then males who mate many times are evolutionarily advantaged over other males, and male promiscuity and aggression in securing mates are expected to evolve, but the same is not true for females. Figure 1.1 shows Bateman's representation of the data from his fruit fly research. He ran two experiments, and in both, the number of offspring increased more dramatically for male fruit flies who mated more often than they did for female fruit flies who mated more often. Bateman says,

> This would explain why ... there is nearly always a combination of an undiscriminating eagerness in the males and a discriminating passivity in the females. Even in a derived monogamous species (e.g. man) this sex difference might be expected to persist as a rule. (365)

And so, biologists widely accept that most male animals, but not female animals, have evolved to be promiscuous and aggressive and that Bateman's principle explains why: it is evolutionary advantageous for males, but not females, to mate many times.

However, it has been argued that this view of the natural world, based on Darwin's ideas from the late nineteenth century, is also infused with Victorian moral sentiment. Jonathan Knight (2002), for example, says that Bateman's work "has been used to extend dated preconceptions about human sexual behaviour to the entire animal kingdom, sometimes to the detriment of scientific knowledge" (254). Knight shows that, while many biologists still endorse Bateman's work as basically sound, most also emphasize that the situation is often much more complicated than Bateman recognized. He notes one biologist's analysis that "the reason Bateman's observation became 'Bateman's principle' is that it appealed to people's intuition about the behaviour of individuals" (256). Figure 1.2 is a much more recent representation of Bateman's principle by Krasnec et al. (2012). Comparison between this figure and Bateman's original figure shows how Bateman's principle has taken on a life of its own. First, the relationship depicted in this treatment is more sharply defined than what Bateman presented based on his data. Second, this treatment posits a relationship to overall fitness, which is a stronger claim than Bateman's finding regarding fertility, just one component of fitness. And so, we find both a codification of Bateman's principle and, increasingly, the recognition that this simple principle does not reflect the full complexity of sexual behavior in animals. A few biologists even reject Bateman's principle entirely, along with Darwin's original observation of aggressive males and coy females, as grounded entirely in scientists' values and expectations instead of sound evidential reasoning (Roughgarden 2004, 2009; Gowaty et al. 2012).

This is one example of how scientific findings can be informed by human expectations. Darwin's observations of the natural world bear a striking resemblance to the social norms of Victorian England, and Bateman's research was later taken to underwrite a general principle in part because it accorded with researchers' intuitions. Research informed by human expectations can still be well founded, and it can still lead to successful results. Darwin's idea of aggressive males and coy females may mirror the expectations for men and women in Victorian England, but it is nonetheless the case that male elephant seals tend to fight one another while female elephant seals do not and that the results of those fights determine male seals' access to mating opportunities. There are also examples of researchers' expectations leading to straightforwardly bad science. Stephen Jay Gould (1996) shows that this is so for research purported to show a genetic basis for IQ

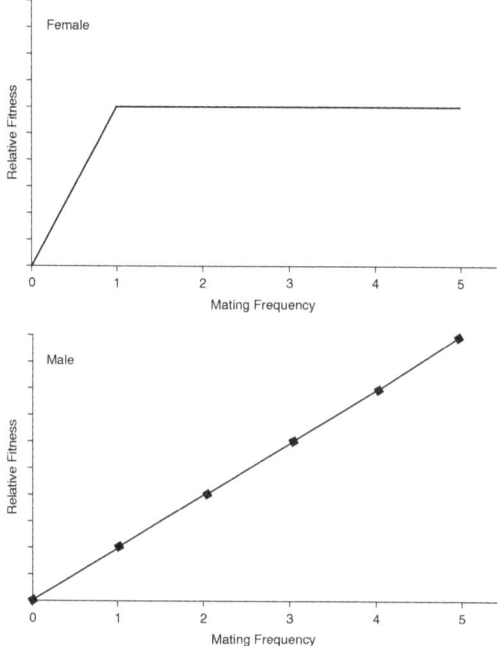

FIGURE 1.2. This figure is a recent illustration of Bateman's principle (Krasnec et al. 2012, Fig. 2), reprinted with permission from Michael D. Breed. Comparison with Figure 1.1 above shows that this presents the principle in a stronger form than warranted by Bateman's original research. First, it represents the difference between males and females as more sharply defined than Bateman's data allowed. Second, it presents the relationship as between number of matings and "relative fitness" rather than relative fertility. This is a stronger conclusion, for it assumes that additional matings have no effect on fitness other than number of offspring produced. This assumption is unlikely to hold true for many animal species.

differences between races, sexes, or social classes. As a second example, Naomi Oreskes and Erik Conway (2011) analyze how a single group of scientists misled the public about tobacco research and environmental issues ranging from acid rain to global climate change, apparently motivated by their political leanings. But even more legitimate research like Bateman's work, as well as scientific results that are even more widely accepted, still bears the mark of human expectations, human concerns, and the limits of human observation.

Features of the natural world that are not anticipated might escape notice entirely. The primatologist Sarah Blaffer Hrdy (1986) shows that data running contrary to the generalization about male promiscuity and female coyness in animals were available many years before any researchers began to question that generalization. In particular, *female* promiscuity is quite common among primates. Hrdy argues that it was thus not newly available data but a shift in the researchers themselves that led to a recognition of the limits of the promiscuity/coyness generalization. She says,

> I seriously question whether it could have been just chance or just historical sequence that caused a small group of primatologists in the 1960s, who happened to be mostly male, to focus on male-male competition and on the number of matings males obtained, while a subsequent group of researchers, including many women (beginning in the 1970s), started to shift the focus to female behaviors. (Hrdy 1986, 159)

Hrdy hypothesizes that this new focus on female behaviors led researchers to finally recognize female promiscuity in primates. This in turn inspired, for the first time, a search for explanations for female promiscuity, instead of merely dismissing those behaviors as anomalies. There is still no agreed-upon evolutionary explanation for female promiscuity. Many biologists think the existing theoretical framework is accurate but needs to be employed differently for cases like the primates (Clutton-Brock 2009). Other biologists think the whole theoretical framework inspired by Darwin's observations of differences between male and female animals needs to be abandoned (Roughgarden 2009). For present purposes, the important point is that, in this case, different individuals with different life experiences performing the science resulted in a significantly altered focus. The changed focus in turn led to an emphasis on different types of interactions and the observation of different patterns.

In philosophy, there is increasing recognition of the myriad ways in which human expectations, concerns, and limitations influence scientific practice. Feminist philosophy of science has attended to this influence for quite some time, including especially the role of social values on the process and products of science. A primary goal of this line of research is to address the question of how social influences on science can be reconciled with scientific objectivity. The current focus in philosophy of science on scientific models, including on the use of abstractions and idealizations, is another area where the limitations of human scientists are considered to be relevant. Wimsatt (2007), for instance, explicitly addresses how idealized models are the result of science produced by limited human agents. Finally, there is a growing literature in philosophy of science that directly addresses the relationship between science and broader social concerns. Special journal issues on the topics of socially relevant and socially engaged philosophy of science (Fehr and Plaisance 2010; Potochnik 2014) address this trend.

The extent of human influences on science makes it especially important for philosophy of science to account for science as it is actually conducted.

The more science is recognized to be the activity of limited human beings, the more its characteristics can be expected to diverge from an idealized rational reconstruction or a hypothesized, perfect end state. Indeed, there is no reason to think that our actual scientific practice and products bear any particularly strong resemblance to a science philosophers would hold to be ideal or to an end state that philosophers would hold to be perfect. As Catherine Elgin (2010) puts the point, "Science is a human achievement—a product of human endeavor. As such, it is ineluctably connected to the ways we access the world" (446). Instead of finding a way to fit our actual scientific practices into a predetermined mold based on antecedent commitments or expectations of science's goals and end point, philosophy of science must take its lead from observations about how science produced by real human beings finds success and what kind of success this is. This is the tack I take throughout this book. Such an approach to philosophy of science is a natural outgrowth of a naturalized epistemology, where our philosophical theories of knowledge are informed by how science in fact proceeds. In this case, the idea is that our philosophical theories of science itself should reflect how science in fact proceeds.

One might have the concern that, if philosophy of science aims to account for actual scientific practice—or *merely* aims to account for actual scientific practice—this would threaten to eliminate any grounds for making normative claims about science. There is, after all, a traditional dichotomy with normative philosophical claims about science on one side and descriptive claims of sociology and history of science on the other.[3] It may seem that providing an account of actual scientific practice lands us on the wrong side of this divide to make any normative claims about science, any prescriptions for how science should proceed. Yet we require a normative account of science in order to articulate epistemically meaningful standards for scientific practice and to distinguish between successful and unsuccessful scientific results in an epistemically relevant way. Limiting ourselves to purely descriptive claims about science would, it seems, undermine philosophy's ability to adjudicate debates about the proper aims, scope, and success of science. It would also in effect eliminate philosophy of science as an enterprise distinct from sociology and history of science.

Happily, aiming to account for science as it is actually practiced does not threaten the possibility of providing a normative account of science. It simply must be a normative account of our *actual* science, as it has in fact grown up, complete with accidental features and limitations due to particular human characteristics. Such an account should be guided by exemplars that are recognized to be our best scientific practices and products, as well as by the standards scientists themselves explicitly embrace or implicitly adopt.

Consider an example of a philosophical position that is at once normative and motivated by ways in which our actual scientific practice deviates from what might be taken to be the ideal. Various feminist philosophers of science have argued that our science fails to satisfy traditional conceptions of objectivity, such as immunity from the influence of scientists' values. Some of these philosophers then use insights about this failure to help motivate an alternative conception of objectivity that they take to be consistent with actual scientific practice. Such revised accounts of objectivity are normative insofar as they outline conditions required to achieve objectivity and place value on that achievement. Consider a specific example of this. Helen Longino (1990, 2001) develops an account of how objectivity emerges from science practiced in certain social conditions, in particular, by a diverse community of scientists with sufficient critical interaction. In her view, this enables the influence of values to be controlled, even though it cannot be eliminated. This may or may not be the right account of objectivity, but it is clearly a normative account. If it is correct, this account has straightforward consequences for how our science should be practiced and what such a science can accomplish. It gives a prescription for good science. Yet subscribing to this normative position requires acknowledging the limitations of our actual science—in this case, how our science falls short of an ideal objectivity. The general point is that normative claims can be made about the actual science that we humans have developed; they simply must be nuanced enough to speak to that enterprise in particular.

There is another concern one might have about a philosophy of science that aims to account only for the actual scientific practices of limited human beings. We might in fact be stuck with our particular science, warts and all, but this does not in itself eliminate any hope of providing a philosophical account of the features *any* science must possess. Such an account would indicate the features that must be possessed by any empirical scientific endeavor, pursued by any beings, in order for that endeavor to be epistemically sound. I said above that there is no reason to think that our science bears a resemblance to any such ideal science, but I have not so far demonstrated that it does not. Nonetheless, I suspect that the interesting generalizations that can be made about science—including generalizations with normative force—are by and large about our human science in particular. The basis for this view is the observation of how deep and thorough the mark of human characteristics is on our science. This can be established by careful attending to our scientific practices, as well as to the epistemic value to which those practices might lay claim.

I thus will move forward with an investigation of our limited, human science in particular with a provisional commitment to this approach. I hope that the nature of my conclusions in this book demonstrates the value of this focus. It may be enlightening to explore the features possessed by any logically possible empirical science and to consider the relationship the features of our science bear to those. But even that project requires a nuanced understanding of the particularities of the science that has been produced by humans (Adrian Currie, in conversation). And I anticipate that many of the significant generalizations to be made about science are distinctive to our science, a science created by humans and tailored to human needs, interests, circumstances, and psychological characteristics. The degree to which other creatures—real or imagined—have a science that mirrors our own depends on the degree to which their needs, interests, circumstances, and psychological characteristics match our own.

The aim of accounting for actual scientific practice seems to be broadly accepted in contemporary philosophy of science. According to de Regt and Dieks (2005), "Nowadays few philosophers of science will contest that they should take account of scientific practice, both past and present. Any general characteristic of actual scientific activity is in principle relevant to the philosophical analysis of science" (139). Similarly, in the introduction to their collection, Chao et al. (2013) state their view that "what really matters to philosophers of science, and what philosophical discussions should be based on, is what scientists actually do and how they do it" (1). Consider just a few examples of philosophers explicitly engaging in this approach. In the preface of his book, Suppes (2002) stresses the importance that "empirical details" have for his account of science. Strevens (2008b) is clear that he intends to provide an account of actual explanatory practice, that is, "what kinds of explanations we give and why we give them" (37). Longino (2001) says that her project is to account for "living science, produced by real, empirical subjects," and she adds that this requires acknowledging that "scientific knowledge cannot be fully understood apart from its deployments in particular material, intellectual, and social contexts" (9).

This emerging philosophical focus on engaging with science as it is actually practiced may in part be responsible for the preponderance of work in current philosophy of science that is intended to generalize only to particular fields of science or even more specific areas of scientific inquiry. For example, the question of reductionism or antireductionism has been investigated as it applies to molecular and classical genetics in particular (Kitcher 1984; Waters 1990). Explanatory strategies have been articulated that are specific to biology (Sterelny 1996) and neuroscience (Machamer et al. 2000). Hierarchical levels of organization have been assessed within physics (Rueger and McGivern 2010) and within macroecology (Potochnik and McGill 2012). Philosophers of science also increasingly engage in debates endemic to particular fields. For example, philosophers of biology have actively participated in debates among biologists about adaptationism (Orzack and Sober 1994), levels of selection (Okasha 2006), and species and phylogenetics (Hull 1978).

Projects in philosophy of science that are tailored to specific scientific fields and subfields are indisputably valuable, but they do not exhaust the possible projects in philosophy of science. One might be tempted to move from a commitment to accounting for actual scientific practice, and the observation that different fields of science are very different in their aims and methods, to the conclusion that any philosophical views about science must be specific to individual fields or subfields of science. I think this would be a mistake. I believe there are still informative generalizations to be made about all of science. These are likely not universal generalizations about methodology, aim, relationship to society, and so on. There is no single, unified scientific method, and the various aims of science include, at least, explanation, prediction, retrodiction, confirmation, generalization, and basis for policy or action. These and other variable features of science are numerous and significant. Yet I suspect that there are also informative generalizations that transcend field boundaries and that apply to a range of scientific projects in other ways quite different from one another (see Currie 2015). This too I expect to be corroborated by the views developed in this book, which are largely intended to be informative generalizations about science as a whole. I return to this point explicitly in the final chapter.

The commitment to account for actual scientific practice is unevenly pursued, even by philosophers who profess it. Here is an example. John Dupré (1993) advocates revisions to classical philosophical positions inspired by actual scientific practice, and one of the positions he criticizes is reductionism, taken as the idea that the understanding of a whole is accomplished by understanding its individual parts. In the process of developing his criticism of reductionism, Dupré declares that the scientific field of population genetics is worthless. Population genetics is a subfield of biology that investigates how evolutionary processes, including natural selection, influence the distribution of genetic variants in a population. Dupré is critical of how population genetics relates the process of natural selection to genes. This is done by estimating the effects different genetic variants have on fitness—that is, how organisms with the different genetic variants fare in survival and reproduction—and then representing these fitness effects with parameters known as genes' "selection coefficients." According to Dupré, this procedure is unacceptable, for it involves the use of "misbegotten mongrel concepts" of genetic fitnesses (138). He objects to considering the fitness of individual genes because this requires applying a higher-level property (fitness) to lower-level entities (genes). For this reason, in Dupré's view, population genetics as a whole is a misguided enterprise.

Dupré (1993) rightly points out that one is not compelled to grant credence to a body of research simply because of the sociological fact that the research is considered scientific. Nonetheless, his conclusion about population genetics is unwarranted. That conclusion is based solely on a philosopher's antecedent expectations for the way in which scientific concepts should reflect reality. Population genetics is an enduring subfield of research. Instead of dismissing this area of research on the basis of a prior philosophical commitment to higher and lower levels and the permissible relationships among them, we would do better to carefully consider the aims of this research and why it persists. We might consider what the scientists in question take themselves to be accomplishing, how that work is received by the scientific community, and, in turn, what philosophical interpretations of its value are available. At the end of this investigation, we may agree with Dupré's conclusion that population genetics is a failed scientific endeavor. But if so, we would be in the position to motivate that conclusion with respect to the specific aims of this field of research. We can "[question] the specific epistemic credentials of particular scientific projects," as Dupré urges (12), but we must do so in a way that is sensitive to scientists' intentions and the reception of their work. This follows from the idea that science is ultimately a human creation and, as such, responsive to particular human concerns.

1.2 Science in a Complex World

The Human Genome Project began in 1990 and was completed in 2003. Its aim was to identify and sequence the roughly 20,000 to 25,000 genes of the human genome. The project was tremendously successful in achieving that aim; it was completed two years early, and the analysis of the results continues to be fruitful. This is an exemplar of big science, spanning 13 years and costing around $3 billion, and with seemingly complete results. Yet the Human Genome Project in fact came nowhere near to providing full information about the human genome. Critics point out that it has failed to identify any genetic causes of diseases, which was one of its primary purposes. For example, 101 genetic variants that were found to correlate with heart disease nonetheless entirely failed to predict the occurrence of heart disease (Paynter et al. 2010). This is sometimes called the problem of missing heritability: the inability to identify any specific genetic influences on traits that are known to be heritable. Missing heritability suggests that the genetic influences on disease are at least much more complicated than anticipated. Perhaps genetic influences also interact with other factors in ways that obscure the causal pathway, or perhaps there is epigenetic inheritance.

Its uselessness in disease diagnosis is just one indication that the Human Genome Project falls short of providing full information about our genes. Consider that human beings are incredibly diverse, with some of this diversity stemming from our genetic diversity. This variation among individual human genomes is wholly unaddressed by the Human Genome Project, which sourced all of the DNA it sequenced from a handful of anonymous donors. Enter the 1000 Genomes Project, the goal of which is to identify most of the common genetic variants in human populations. Instead of treating the human genome as if it were a single, uniform entity, the 1000 Genomes Project directly addresses human genetic diversity. There is certainly great value to "mapping" the human genome. This identifies what regions of our DNA are genes, that is, are directly used to make proteins, which is the most direct path of genetic influence. But the variation in the genetic material occupying some of those regions is immense. The 1000 Genomes Project directs attention to determining what the most common variations are.

Even the scope of the 1000 Genomes Project is limited, though. According to the project's website, its samples are drawn from only 27 human populations, and it is designed to only capture the most common variants, those possessed by more than 1% of a population. So the full extent of human genetic diversity is greater than this project can identify. Both of these enormous projects in genomic research, involving collaborations among large numbers of researchers and institutions across the world, have yielded a great deal of information and insight. Yet both still fall far short of providing a full account of human genomes. Moreover, this is just the tip of the iceberg when it comes to the complexity of the genetic influences on humans. Microbial cells actually outnumber human cells in the human body: the majority of your cells are not human cells. And these microbes are not just along for the ride. The makeup of microbial cells strongly influences human development and physiology. The Human Microbiome Project is yet another massive project in genomic research. It uses DNA sequencing to provide a "metagenomic" analysis of the microbial communities of human beings. In other words, this project aims to analyze the diverse microbial genes—genes that are not human but that also occupy human bodies and that influence human traits much as our own genes do. Figure 1.3 shows images representing these three massive projects in genomics and summarizes how each expands on the last.

These three massive, collaborative projects of modern genomics have incredibly ambitious aims; indeed, each aims for the completeness of some

INTRODUCTION

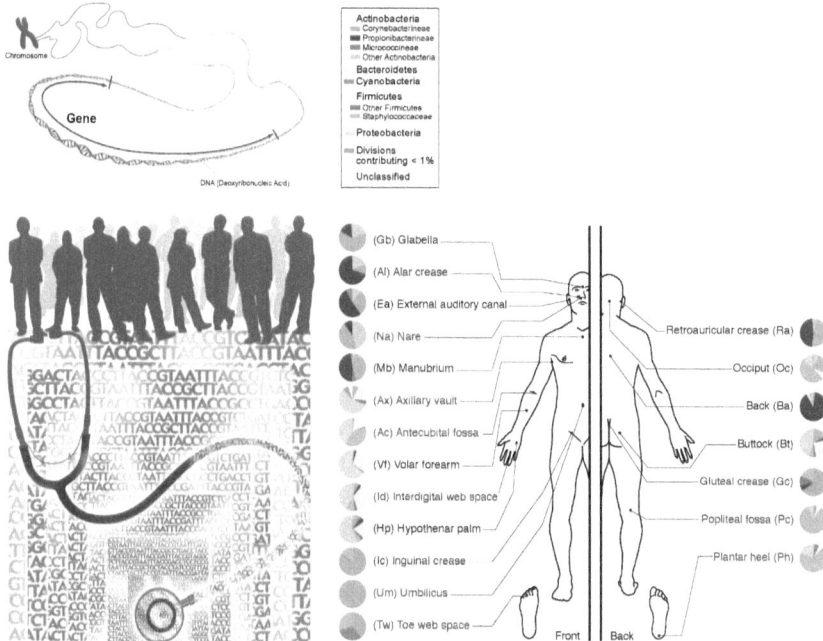

FIGURE 1.3. *Top left*: The Human Genome Project resulted in a full sequencing of all 23,000 human genes. Credit: Darryl Leja, NHGRI (www.genome.gov). *Bottom left*: The 1000 Genomes Project examined DNA from members of 27 human populations and identified 16 million previously unknown variations in those 23,000 human genes. Credit: Jane Ades, NHGRI (www.genome.gov). *Right*: The Human Microbiome Project aims to sequence the genomes of the numerous microbes on and in the human body; these genomes contain over one million genes in all, many of which influence human traits. If the 23,000 human genes have more than 16 million variants, imagine the extent of variation contained in our microbiome's millions of genes. Credit: Darryl Leja, NHGRI (www.genome.gov).

body of information. And yet, each also clearly falls short of providing a complete accounting of the genetic influences on human traits, as evidenced by the need for the other projects. Furthermore, it is one thing to sequence these genomes, and it is quite another to tease out which genes influence which traits, not to mention *how* those genes exert their influence. The latter project is pursued in other areas of biological research. Finally, all three of these projects together—the Human Genome Project, 1000 Genomes Project, and Human Microbiome Project—leave entirely untouched the broad range of nongenetic influences on human traits. These include, at least, direct environmental influence and environmentally mediated gene expression, the evolution of genetic influences, the evolution of developmental and environmental mediation of gene expression, and, in some cases, learning or social influence. All of these types of genetic and nongenetic influence on traits are listed in Table 1.1, along with areas of research in which they are investigated.

TABLE 1.1. Despite their ambition, these genomics projects ignore many influences on human traits. Here is a list of some of those influences, along with other areas of research in which they are investigated.

Influences on human traits	Areas of research investigating them
DNA sequencing, mapping of molecular genes	Genomics
Which genes influence which traits	Classical and molecular genetics
How genes exert influences on traits	Developmental biology
Environmental mediation of gene expression	Phenotypic plasticity
Evolution of genetic influences	Evolutionary biology
Evolution of developmental/environmental mediation	Evolutionary developmental biology
Learning or social influence	Behavioral ecology, social sciences

The enormous complexity of the causal influences on human traits is thus evident, as are the scientific difficulties created by grappling with all of this complexity. And the causal influences on human traits are just one small pocket of all the fascinating and complicated happenings in this wide world.

It's a complex world. The characteristics of all living organisms, including us humans, result from exceedingly complex processes of genetic influence, evolution, developmental factors, and direct environmental influences. Those characteristics exist in seemingly endless variety. The social lives of most animals are similarly diverse and are shaped by an even more extensive range of influences. Events in inorganic systems are similarly complex. Climate change arises from a diverse range of human and nonhuman influences, and its effects on weather, wildlife, human agriculture, and economies are numerous and unpredictable. Even apparently simple physical phenomena—like a ball being hurled through a glass window or water dripping from a faucet—are surprisingly complex upon deeper investigation.

In this quick description, I have gestured toward several ways in which our world is complex. First, there is simply a large variety of different phenomena, for example, the seemingly endless variety of traits exhibited by living organisms. Second, there is an extensive range of influences on any phenomenon, for example, the numerous influences on any given trait of an organism (see Table 1.1). Those influences are also phenomena in their own right, and they too vary immensely. Third, the influences on similar phenomena vary and also combine in different ways. For example, different traits of a single organism are shaped by different influences, as are the traits of different organisms, of the same species and of other species. Finally, there is even complexity in how individual influences affect a phenomenon. Causal interaction, feedback loops, and nonlinear effects of various types abound. Our world is not simply complex; it is multiply complex. This raises the question of whether what

I'm calling "complexity" is just a grab bag of unrelated observations about our world. In a sense, this may be so. There are a variety of different aspects of the world that are more complex than it seems they might have been, and those aspects may be to some degree independent of one another. On the other hand, I expect that this set of complex aspects of the world is similar in an important way. I'm not yet in a position to develop that idea, so I postpone further discussion of the forms of complexity until the next chapter.

One also might wonder if positing a complex world and the idea that human limitations and concerns influence science aren't just two sides of the same coin. Complex phenomena may in part be considered complex in virtue of their surprisingness to us. If we were otherwise, without our particular sensory and cognitive propensities and limitations, we may not be so surprised. A second reason to connect human influence on science to the complexity of phenomena is that these complexities exist among features of the world we choose to focus on and in the accounts we formulate of those features. Perhaps the complexity is only apparent; perhaps if our science proceeded in a different way, the complexity would diminish. This brings to mind watershed transitions in the history of science that increased simplicity and predictability and decreased the appearance of complexity, such as shifting from a geocentric to heliocentric model of the world. There may well be connections between the human influences on science and our science continually uncovering complexity. But the former does not entail the latter. In the history of our specifically human science, we might have encountered less variety in phenomena and their influences and less complexity in how those influences are exerted. The observation that our science encounters a complex world thus goes beyond the observation that our science bears the marks of its human practitioners.

In light of the extent and variety of complex phenomena, it should be no surprise that approaches designed to accommodate complexity are employed in a number of fields of science. Dynamical systems theory applies differential or difference equations to the behavior of a wide range of complex, dynamical systems (Chemero 2011). One example of an application is to the development of human motor skills. Developmental systems theory grapples with the interactions among genetic, epigenetic, and environmental contributions to development in biological organisms (Oyama et al. 2001). Systems dynamics uses feedback loops to represent nonlinear behaviors, while chaos theory is applicable to systems that are highly sensitive to initial conditions. Systems biology examines molecular action in a broader biological context (Kitano 2002). These are only a few examples of the many scientific applications of approaches developed to grapple with complex phenomena.

The prevalence of complex phenomena has also been appreciated by a range of philosophers, many of whom have gone on to consider the implications of this complexity for science. Cartwright (1983) shows how inaccurate our fundamental laws of physics are of the messy phenomena to which they are applied. Dupré (1993) argues for a range of metaphysical conclusions that he thinks arise from the disorderedness of the phenomena under scientific investigation. Cartwright (1999), in turn, offers a vision of the lack of unity among the findings of different fields and subfields of science based in part on the untidiness of phenomena—that is, the "dappled world." Mitchell (2003) discusses the complexity of biological phenomena in particular, which involve multiple components at multiple levels, while Mitchell (2012b) focuses on the implications of such complexity for our scientific understanding and the policy decisions it informs. Strevens (2006) considers how it is that simple patterns can be observed in complex systems with many strongly interacting components.

Yet the full implications of this complexity are even more striking than is generally recognized by scientists and philosophers of science. Many scientists, deeply enmeshed in their particular research agenda, do not sufficiently appreciate how many features of the phenomenon under investigation are overlooked by their approach or how many influences on that phenomenon their approach neglects. It does not follow from this that these scientists' methods are flawed, nor does this undermine the results of their investigations. As I urged in the case of Dupré's criticisms of population genetics, we must carefully consider what scientists take themselves to be accomplishing, how that work is received by others in the scientific community, and what philosophical interpretations we might offer of its value. Together, these provide some insight into the nature and limitations of research projects that tackle complex phenomena from some specific angle. If I am right that complexity is pervasive and that many scientists overlook some limitations of their research, then we should not take at face value scientists' claims to have fully addressed some phenomenon. We should instead expect there to be other research, by other scientists and utilizing different methods, that provides its own account of the same phenomenon. Overlooking limitations of their research can also lead scientists to err on the side of overemphasizing conflicts between their research findings and others'. If research projects have subtly different goals, as do the projects of genomics surveyed above, then this might lead the approaches to emphasize different features of a phenomenon. Findings that apparently conflict may in fact illuminate different aspects of a complex phenomenon.

As it turns out, this is common. In Potochnik (2013), I discuss two apparently ideological rifts in population biology, that is, methods or theories to which some biologists are fully committed and others are fully opposed. I argue that these disagreements result from harmless differences in approach, differences that we should expect in light of the complex phenomena these biologists are investigating. Longino (2013) provides case studies of numerous different approaches to studying human aggression and sexuality. She finds that, although the approaches are often taken to provide competing accounts of these behaviors, they actually have subtly different aims from one another. Somewhat similar to the genomic projects considered above, the approaches vary in part as to whether their focus is the trait itself, variation among individuals, or variation among populations. These different focuses lead to an emphasis on different types of causal influences. And so, according to Longino, instead of providing competing accounts of a single phenomenon, each of these approaches investigates just a few of the many types of causal influences on human aggression and sexuality. In Chapter 3, I will analyze some of the approaches to human aggression that Longino addresses and the relationships among them. For now, the point is that losing sight of the full extent of complexity can lead scientists to overlook the limitations of their research and overemphasize conflicts with other research approaches.

Philosophers too can fail to recognize the full implications of the world's complexity for science. Many philosophical accounts of science's aims and methods, and of what science actually shows us about our world, do not adequately reflect how the features of our science are shaped by the need to grapple with complex phenomena. Consider Michael Strevens's (2008b) kairetic account of scientific explanation. Strevens numbers among those who are impressed by our world's complexity, and he is also explicit that his aim is to account for actual explanatory practice. Even so, Strevens's account requires that any full scientific explanation must "black-box" nothing, that is, must include no functional specifications. He grants that explanations are most effectively pursued piecemeal, with each part black-boxing a range of difference-makers. Nonetheless, he requires that any *full* explanation bring these partial explanations together into "a single whole" (161). Complexity renders this requirement impossibly strong even as a programmatic goal. Assumptions that black-box some causal influences occur in virtually all scientific representations and for good reason. As the genomics example illustrates, any phenomenon is shaped by a tremendous variety of influences. Simplifying assumptions are needed in order to successfully represent only a handful of those influences. Combining representations of disparate influences into a single, unified representation is oftentimes impossible, if only because of the limitations of our modeling techniques and computational powers. This impossibility is made apparent, I think, when Strevens clarifies that "you may have to visit many other university departments, finishing of course with the physics department" to obtain a full explanation (161). In our complex world, visits to so many scientists engaged in different pursuits would not culminate in a full explanation of anything. At best, this would yield a hodgepodge of partially related representations, with no way to reconcile their incompatible assumptions and no way to combine their mathematical frameworks. I hope the conclusions of this book will demonstrate that the influence of complexity on science is not so easily dismissed or set aside. Indeed, its significance is difficult to overstate.

1.3 The Payoff: Idealizations and Many Aims

These two ideas serve as my starting point for this project. First, science is practiced by limited human agents, and the main job of philosophers of science is to make sense of the actual scientific practices that we humans have developed. Second, we inhabit an exceedingly complex world, and this is reflected in our scientific practices. I have motivated these two ideas in this chapter. I do not take myself to have conclusively demonstrated either of these ideas; they are simply supposed to be plausible initial assumptions. The first—that philosophy should account for science as humans have in fact developed it—is a view about the purpose of philosophy of science that informs the aims of my project. The second idea—that scientific practice is shaped by our world's complexity—is a substantive claim about science. This claim will be further corroborated as my project proceeds. I have discussed how versions of these two ideas are accepted by many philosophers of science. Nonetheless, a thoroughgoing commitment to these ideas inspires some novel philosophical positions. In the remainder of this chapter, I preview those positions and indicate how they are developed in the book to come.

The most basic consequence of science practiced by humans in a complex world is the expansive use of idealizations, or assumptions made without regard for whether they are true and often with full knowledge that they are false. In Chapter 2, I make the case that science involves pervasive and ineliminable idealizations, and I diagnose the reasons for this situation. I begin with the observation that universal, exceptionless laws are few and far between in science, if there are any at all. Scientific laws, theories, and models typically only hold approximately, in some range of circumstances, and they liberally employ idealizations to accomplish this partial fit. I argue that they depict *causal patterns*. These patterns qualify as causal according to a manipulability approach to causation. Causal patterns are limited in two ways: they hold only over some range of circumstances, and they tend to have exceptions even within that range. Science's focus on causal patterns is due to this complex world we inhabit. Armed with a manipulationist approach to causation, I can more carefully characterize the relevant sense of complexity, which I term *causal complexity*. Uncovering causal patterns is difficult in the face of causal complexity, and a ubiquitous method of overcoming these difficulties is the use of idealizations. The second half of Chapter 2 is devoted to providing an account of the status and purpose of idealizations. I defend the idea that there are many *intertwined reasons to idealize* that reflect not just features of the world but also researchers' interests. I also argue that idealizations play a positive representational role: they *represent as-if*, that is, represent systems as if those systems possessed features they do not. For these reasons, idealizations are widespread. There also tends to be little focus on eliminating them in favor of more accurate representation or even on controlling their influence. Idealizations are thus *rampant and unchecked* in science. This observation is the linchpin of all the ideas about the aims of science defended in subsequent chapters.

In Chapter 3, I move from generalizations about all of science—the significance of causal patterns and the importance of idealization—to an examination of the ways in which scientific research varies. I do this by surveying a handful of different scientific projects. Behavioral ecology research into the evolution of cooperation is a clear illustration of how causal patterns can be sought without much attention to any individual phenomenon. A variety of investigations into human aggression demonstrates the changed significance of causal patterns when the research focus is a specific phenomenon, with an eye toward intervention. Finally, a more superficial investigation of projects in the physical sciences, including fluid dynamics, quantum physics, and climate science, reveals additional variation. All of these research projects vary in their levels of abstraction, their relationships to data, and their aims of representation, prediction, or intervention. Nonetheless, all crucially employ idealizations, largely without any steps taken to minimize them or control their influence. These diverse scientific projects thus corroborate the idea that idealization is rampant and unchecked in science.

These case studies position me to launch into an analysis of the very aims of science in Chapter 4. With the centrality of idealizations in science further confirmed, I suggest that the distance between idealized representations and traditional articulations of the aims of science should lead us to question those traditional articulations of science's aims. I argue that the epistemic aim of science should be taken to be not truth but *understanding*. Nothing has gone wrong with or is lacking from idealized representations, and no intermediary step is needed for them to generate human understanding. Understanding has a dual nature: it is an epistemic achievement, but it is also a human cognitive state. I follow Catherine Elgin (2004, 2007) in holding that sometimes understanding can be advanced by departing from the truth, including with the use of idealizations. The account of scientific understanding I develop permits greater and more variable departures from truth than does Elgin's, departures that are motivated in part by scientists' immediate research interests. In my view, this position is largely orthogonal to the debate between realism and antirealism, but it does require reconsideration of the nature of scientific knowledge. Science also has a host of other aims besides understanding, and I argue that successful pursuit of one of these aims often occurs at the expense of others. This helps account for the variety of projects found in science, as well as continuing disagreement among scientists about the value of these various projects. A consequence of these views about the epistemic aim of science and its relationship to other aims is that the products of science are for the most part not things we believe to be true.

Scientific understanding is generated via the production of scientific explanations. Successful explanation explicitly depends on features of human psychology and cognition as much as it depends on features of the world. Yet this observation has had remarkably little influence on philosophical accounts of scientific explanation. In Chapter 5, I develop a causal pattern account of explanation that reflects how our explanations are shaped by their communicative purposes. The ontic dimension of explanations, the explanatory form of metaphysical dependence, is causal pattern. Information about causal patterns—about causal dependence along with the scope of dependence—is what explains. This is due to causal patterns' special contribution to human understanding. Any given phenomenon embodies many causal patterns, and the particular research program under way is crucial to determining which of these causal patterns explains, that is, which results in understanding. Finally, there is a general *criterion of explanatory adequacy* that any explanation must satisfy. This is an entailment requirement patterned off similar criteria others have proposed. But my version of the requirement is much weaker and more variable in virtue of the fact that idealizations contribute to its satisfaction.

Chapter 6 is an investigation of the concept of hierarchical levels of organization and various forms of significance that have been attributed to levels. The idea that our science reflects well-defined levels is based on the simple idea of composition. However, although the relationship of composition may be straightforward and universal, discrete levels are not—at least not among

INTRODUCTION

the phenomena targeted in scientific investigation. Attention to the notion of causal complexity, the strategy of continuing idealization, and science's wide array of projects can help separate the straightforward, plausible claims about composition and size ordering from a range of dubious ideas about levels they often have been taken to imply. The result is a thoroughgoing rejection of universal hierarchical ordering, stratification into levels, and the scientific relevance of realization, including multiple realization. Classically, the fields of science have been taken to reflect levels of organization. I argue, against this, that the metaphysical significance of field divisions is quite limited. The fields of science simply reflect the fact that phenomena are complex and only ever addressed partially with highly simplified approaches. This results in many different approaches, spanning subfields and fields, that resist integration. As for the question of the relationships among these fields and subfields—classically, the issue of the unity or disunity of science—the same combination of causal complexity and idealization that necessitates field divisions also motivates cross-fertilization among the different approaches that result.

A significant implication of the view of science's aims developed in this book is an expanded space for the influence of human characteristics on the practice of science, including our social values. I explore this implication in Chapter 7. The conception of the aims of science I develop in this book loosens the relationship between successful science and truth, and this creates avenues for researchers' priorities and concerns to directly influence scientific findings. Ideas about how social values find their way into science are often linked to conceptions of scientific pluralism. My view diverges somewhat from this pattern, for the same features of science that ground an expanded role for social and political values also act as a brake on pluralism. This includes the idea that there is a more distant relationship between science and metaphysics than is sometimes conceived, so divergent scientific results have little metaphysical import. The articulation of the role for values in science I develop also leaves intact and, indeed, enhances our ability to assess divergent scientific approaches. Finally, this project involves a number of generalizations about science as a whole, including the prominent role of causal patterns, the continuance and significance of widespread idealization, how explanations proceed, and the relationships among fields. These generalizations conflict with versions of pluralism that anticipate few or no commonalities across scientific endeavors. I conclude what is largely a highly abstract, philosophical treatment of science with a discussion of the implications this treatment has for actual scientific practice.

Before embarking on this project, let me forestall a concern with the two starting points outlined in this chapter and the views I will develop on their basis. The account of the aims of science put forward in this book should not be viewed as disparaging science's promise and success. My project is not to criticize the success of science or to undermine its well-foundedness. To the contrary, I believe that the first step in appreciating the great success of our scientific enterprise is to accurately characterize its intent. Science is a human tool. It is a remarkably powerful tool, and it is surely our most important epistemic tool. Once we properly appreciate the aims of science and the contexts in which those aims are pursued, features of science that appear to be shortcomings are instead revealed to be strengths.

2

Complex Causality and Simplified Representation

In this chapter, I establish the most basic consequence of science performed by humans in a complex world, namely, widespread idealization. As a first step, I make the case that much of science is profitably understood as the search for causal patterns. By and large, the causal patterns discovered hold over a limited range of circumstances and have exceptions even within that range. Moreover, whether a causal pattern emerges depends on our representational choices. I employ a manipulability approach to causation to articulate the nature of causal patterns, but I remain agnostic on the status of manipulability relations in the metaphysics of causation. I do, however, defend the idea that manipulability relations are epistemically basic in causal analysis. This discussion of causation and causal patterns grounds a more specific characterization of the relevant sense in which the world is complex, which I term "causal complexity." I argue that causal complexity is a pervasive feature of the world that significantly affects scientific practice.

In the second half of the chapter, my attention turns from the scientific significance of causal patterns and causal complexity to a prominent scientific strategy for discerning patterns in the face of complexity, namely, the use of idealizations. I argue that idealization plays a more expansive role in science than is commonly appreciated by scientists or philosophers of science. Idealizations are widespread even outside model-based science, and there are many different, intertwined reasons to idealize. Moreover, I motivate the idea that idealizations actually play a direct representational role. As a result, idealizations are both rampant and unchecked in science. Exactly what I mean by this will become clear below.

2.1 Causal Patterns in the Face of Complexity

2.1.1 CAUSAL PATTERNS

In Chapter 1, I discussed the shift in philosophy away from accounting for an eventual, perfected science or a prettified, more rational version of science toward instead accounting for today's actual scientific practice, messiness and all. This shift has been accompanied by a great deal of attention to ways in which science falls short of discovering laws, taken as exceptionless regularities that are universal in their scope. One influential critic of universal laws is Nancy Cartwright (1983), who argues that physics' most fundamental laws are in fact false. In Cartwright's view, those laws are successful, for they are explanatorily powerful, but they do not accurately describe nearly any real systems. Cartwright gives the example of Newton's law of universal gravitation, which states that any two bodies attract each other with a force directly proportional to the product of their masses and inversely proportional to the square of the distance between them:

$$F = G \frac{m_1 m_2}{r^2}.$$

Here F is the force, G the gravitational constant, m_1 and m_2 the two masses, and r the distance between the centers of those masses. As Cartwright points out, the physicist Richard Feynman (1967) called this law "the greatest generalization achieved by the human mind" (14). Yet, to accurately capture the behavior of two bodies, the law of universal gravitation requires the assumption that gravity is the only force acting on those bodies. This means that the law does not accurately capture some cases, such as when bodies are charged, so also subject to electrical force. The law also assumes that the mass of bodies is concentrated in a single point, that is, it applies only to "point masses." This assumption is, of course, false of every body. It does accurately reflect the *behavior* of some spatially extended masses, but this is so only for bodies that are spherically symmetrical. Cartwright argues that all of our most illuminating laws have these kinds of shortcomings. Accordingly, on her view, it is not that our scientific laws occasionally have exceptions but that they do not hold true for virtually any real systems. They are, strictly speaking, false.

Considerations like those Cartwright introduces have resulted in philosophical investigations of science all but abandoning the conception of laws as universal, exceptionless regularities. Ronald Giere (1999) also argues that so-called scientific laws are not true. However, rather than maintain with Cartwright that the nature of these laws must be reinterpreted, Giere instead concludes that there simply are no laws of nature. Other philosophers have developed alternative conceptions of laws that permit limitations and exceptions, and still others have investigated the status of ceteris paribus laws. Additionally, increasing philosophical attention has been directed toward other products of science besides laws, including models, causal analyses, and mechanism sketches. The covering-law approach to explanation due to Hempel and Oppenheim (1948) has been replaced by a range of other approaches. Machamer et al. (2000), in a seminal paper in the mechanisms literature, note that laws of nature have little if any relevance to their fields of focus, which are neurobiology and molecular biology. All of this illustrates that there is increasingly an acknowledged diversity of scientific projects, and although opinions vary on the centrality of laws, universal, exceptionless regularities are taken to be uncommon or entirely absent.

There is, however, a different characterization that better describes these diverse scientific projects. A central aim of science is to uncover *causal patterns*. Or so I will argue. Let me first discuss what I mean by "patterns" before moving on to address the sense in which these patterns are causal. I have spoken as if, classically, laws were supposed to be universal regularities in phenomena, about which we humans make generalizations. There is sometimes ambiguity about which is the law—the regularity or the generalization made about it. Such ambiguity is of little consequence if science accurately depicts (or provides) universal, exceptionless laws of nature. It becomes problematic, though, when that construal is questioned and the truth or reality of laws along with it. If science is supposed to provide insight into universal, exceptionless laws of nature, then following Giere, those laws are not real; if scientific laws are our universal, exceptionless generalizations about the world, then following Cartwright, those laws are not true.[1] I intend causal patterns to be regularities in phenomena themselves. On this view, patterns are not themselves human representations but are depicted by our representations. Conceiving of causal patterns as regularities in phenomena—and thus as *real*—turns out to be important for my account, as will become clear in Chapter 4.

And yet, as the term suggests, patterns are not universal and exceptionless but limited in scope and permit deviations and exceptions. This accords with the primary themes of the various criticisms and revisions that have been made of the classic concept of laws of nature. For example, Giere (1999) suggests that what had traditionally been understood as universal laws are instead "restricted generalizations." Like Cartwright, Giere also uses the example of Newton's law of universal gravitation. He points out that to productively employ this law, one must explicitly restrict it to certain kinds of systems, and even then, one must often employ approximation techniques before the systems appear to embody the posited relationship. Consider first the idea that generalizations like the law of universal gravitation have a restricted domain of application. This is because the patterns they depict hold only in a limited range of circumstances. Consider the ideal gas law, $PV = nRT$, where P is the pressure of a gas, V its volume, n its number of molecules, T its temperature, and R the ideal gas constant. The pattern represented in this generalization only holds in a limited range of circumstances; it does not obtain at very low temperatures or at very high pressures. Appending a ceteris paribus clause is sometimes used to acknowledge this kind of limitation, especially when the exact circumstances in which a pattern obtains are unclear. Now consider the second idea, that generalizations can also have limited accuracy even within their domain of application. This is due to deviations from the patterns depicted by generalizations. The ideal gas law illustrates this too, for it ignores both molecular size and intermolecular attraction, and these factors can diminish its accuracy. It is possible to introduce correction terms, which yields the van der Waals equation, but the original form of the ideal gas law is often judged to be sufficient or even preferable. When deviations from a pattern are great enough, this results in exceptions to the pattern, in other words, systems that we would expect to embody a pattern failing to do so.[2]

The patterns identified in science are, then, limited in two ways: they hold only in limited circumstances, and most also have deviations and exceptions even within those circumstances. This is unsurprising once one takes seriously that our science is the product of limited human faculties and concerns, grappling with a world as complex as ours. The simplicity and straightforwardness of a pattern at once increase its usefulness for limited humans and decrease its universality in a world that is neither simple nor straightforward. This is so for fundamental physics as much as for biology and sociology. Physics is the product of limited human beings as much as any other field of science, and physical phenomena outside the laboratory are just as complex.[3] Powerful generalizations such as Newton's law of gravitation would be lost if science sought only universal and exceptionless regularities. And such generalizations are members of a broader class of scientific products, a class that also includes generalizations with explicit ceteris paribus clauses, models, mechanism sketches, and many more types of what I will broadly refer to as representations. The members of this class all depict patterns—limited patterns with exceptions.

This notion of patterns can be further elucidated by considering what Daniel Dennett (1991) calls "real patterns." Dennett's main focus in that paper is not central to the current discussion, but an adapted version of his treatment of patterns yields considerable insight here. One change I will make is

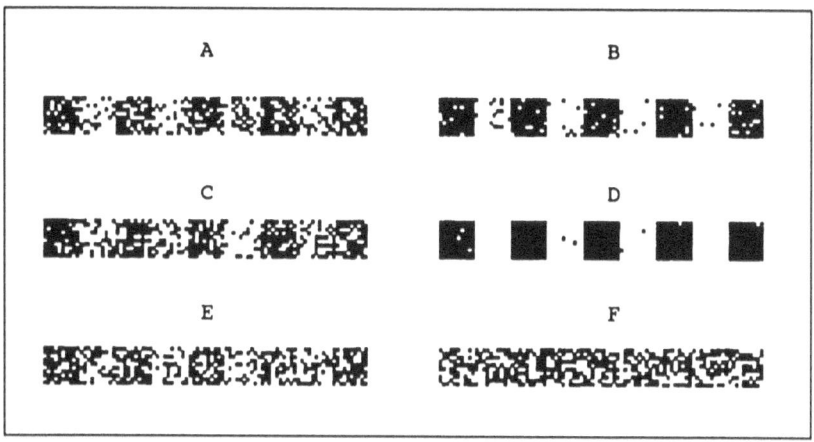

FIGURE 2.1. Dennett (1991) uses these objects to articulate his conception of a pattern. A pattern that Dennett calls "bar code" is embodied by A through E, although each deviates from the pattern to some degree. If the deviation is great enough, this leads to an exception to the pattern, as in F. Reprinted with permission from Daniel Dennett and the *Journal of Philosophy*, copyright 1991.

that, where Dennett considers patterns in data, I will instead consider when a pattern is exhibited by phenomena. Put another way, my question is what it means for a phenomenon to *embody* a pattern. This is because I intend causal patterns to be regularities exhibited by phenomena. Consider Figure 2.1. This is Dennett's simple example of a pattern, which he dubs "bar code." Bar code is the pattern of 10 rows of 90 dots each, with every row a repetition of 10 black dots and then 10 white dots. Take these instances of bar code to together comprise a set of phenomena. Notice, first, that the pattern bar code is present with different degrees of noise, or deviations from the pattern. In these six phenomena, the level of deviation from bar code varies from 1% to 50% of the dots comprising each object (D and F, respectively). Despite this variation, A through E embody the pattern bar code. The pattern is most perfectly embodied by D, but if you squint, you can just make out the same pattern of dots in E. On the other hand, F, with 50% deviation, has the same appearance as a random 90 × 10 matrix of black and white dots.

One might exactly depict any one of these phenomena by first describing or referencing the pattern bar code and then detailing the individual dots that vary from that algorithm. However, as Dennett (1991) points out, sometimes full information about the variance is unnecessary, in which case one might depict a system by citing the pattern bar code and indicating the level of deviation from that pattern. This yields an identical description for A and C, each of which exhibits 25% deviation. Notice that this decrease in the specificity of description can be continued. Sometimes we may not even want to indicate the amount of deviation; perhaps all that matters to us is whether a system embodies the pattern in question. In such circumstances, A through E have the same description: all simply are bar code.

Our so-called scientific laws tend to be like that. (Again, let's set aside the question of whether they deserve to be called laws.) Each depicts a pattern instantiated by a range of systems without indicating how much deviation we should expect or that there might be exceptions.[4] A law presents itself as universal by not containing information about the limitations of the pattern it depicts. As Cartwright (1983) emphasizes, some of the value of our scientific laws derives from this feature. This is also why approximation techniques are often required to accommodate a system's deviations from the pattern. Mistakenly construing these patterns as universal, exceptionless laws of nature occurs when, instead of ignoring the deviations, it is posited that there are in fact no deviations or that any deviations are insignificant.

Consider again system F in Figure 2.1. F does not embody the pattern bar code. Dennett (1991) points out that the pattern is indiscernible in F and suggests that the idea of an indiscernible pattern is self-contradictory. But notice that there is a related pattern that *is* embodied by F, as well as by all the other systems in the figure. A through F are all produced by the same algorithm, namely: produce a dot matrix of pattern bar code, but change the color of a specified number of randomly selected dots. Call this bc-algorithm. Simply looking at the system F will not reveal that it embodies pattern bc-algorithm, but witnessing its production or reading Dennett's account of its production does. This indicates another important feature of patterns. Systems can embody multiple patterns. System A embodies the pattern of bar code with 22 different dots (specific, although not here specified), A and C both embody the pattern of bar code with 25% deviation, A through E embody the pattern bar code, and all systems A through F embody the pattern bc-algorithm. System A thus embodies at least four patterns. The phenomena confronted in science also embody multiple patterns. Those patterns may be embodied with different degrees of deviation, as the simple example of system A also illustrates. Similarly, a gas may embody the pattern depicted in the ideal gas law, even if the van der Waals equation is more accurate of its behavior.

Notice three additional features of how phenomena embody multiple patterns. First, those patterns may involve different features of a phenomenon. Whether a dot matrix embodies the pattern bar code depends only on its appearance, whereas a dot matrix embodying bc-algorithm depends only on its process of creation. Accordingly, whether a pattern is salient depends on what we can and choose to observe about a system, as well as what we can and choose to represent about the system. Second, notice that this feature of patterns enables human characteristics and interests to exert an influence on the salience of a pattern, as well as on which patterns receive our focus. Dennett (1991) points out that two individuals may perceive different patterns in the same data, in virtue of the individuals' different interests and perspectives. So too might different individuals, with different interests and perspectives, judge a system to embody different patterns. Whether the ideal gas law or the van der Waals equation is better in some circumstance depends not only on features of a gas but also on the nature of our interest in the gas. Third, although phenomena embody many patterns, perhaps even countlessly many, it is not the case that phenomena embody every pattern. There is a fact of the matter about whether a phenomenon embodies some pattern.

Laws of nature were originally supposed to be universal, exceptionless regularities. Patterns are neither universal nor exceptionless, but they are regularities. In particular, unlike the simple bar code example, the patterns sought in science are regularities in dependence relations. The law of universal gravitation shows how the force between two bodies depends on their mass and the distance between them. The ideal gas law shows the interdependence among a gas's pressure, volume, and temperature and how these depend on the amount of gas. This focus on dependence relations is at least in part due to science's focus on facilitating human action and human comprehension. Uncovering predictable relationships of dependence is key to both of these goals. A mastery of dependence relations—of how a shift in one thing changes another—renders the world's behavior and the effects of our own actions more predictable. These patterns of dependence are, by and large if not without exception, *causal* patterns.

I will motivate this idea by appealing to James Woodward's (2003) manipulability approach to causation. A manipulability approach is appropriate for my purposes for two reasons. First, it puts human action at center stage. The practical utility of causal knowledge in manipulation and control is a key motivation for Woodward's account.[5] This strategy of grounding causal analysis on action, and specifically human action, is appropriate for a science that is the product of humans. We should expect the sort of causal information uncovered by science ultimately to derive from changes in the objects and types of objects observed by humans. Second, for Woodward, causal relationships among types are fundamental and the basis for causal attributions in singular instances. Causation is taken to ultimately be a relationship among variables, where variables are "properties or magnitudes ... capable of taking more than one value" (Woodward 2003, 39). Woodward is, I think, persuasive that this reflects how scientists themselves use causal reasoning. It also reflects what I take to be the pervasive focus on generalization across multiple instances—that is, a focus on patterns.[6]

These two advantages of Woodward's (2003) manipulability account of causation are reflected in the two core concepts of that account: intervention and invariance. Suppose you want to ascertain whether one variable, call it X, causally influences another variable Y. An *intervention* on X changes its value surgically, so to speak, so that any change induced in Y is due solely to the change to X, not to any side effects of the intervention. Woodward likens his concept of intervention to an "ideal experimental manipulation." For Woodward, X is a cause of Y just in case some interventions on X change Y's value in some background circumstances. Interventions, or manipulability relations, are thus the ultimate guide to causal relations. Then, if X is a cause of Y, that causal relationship is *invariant* over some interventions and range of background circumstances. That is, the causal relationship continues to hold in those conditions. Causal relationships all require some amount of invariance, for how changes to X affect Y must at least be stable under *some* intervention in some circumstance. But invariance comes in degrees. A causal relationship may be stable over larger or smaller ranges of interventions and background circumstances. Invariance is key to formulating generalizations about causal relationships.

Most, and perhaps all, patterns of dependence uncovered in science qualify as causal on this approach. It's broadly accepted that much of science trades in causal information, so I will focus on the potentially problematic cases, patterns that seem like they may not be causal or may not be causal according to a manipulability approach. Notice first that Woodward's (2003) account does not require an actual intervention to be performed in order for a causal relationship to exist or for a causal relationship to be ascertained. The connection between intervention and causation is instead conceptual. Woodward claims that causal relationships can exist even when an intervention is not physically possible, so long as a surgical change to a variable's value is at least physically possible by other means. Thus, although rooted in facts about manipulability, the causal patterns ascertained in science can extend beyond the directly manipulable. This is as it should be, for science's reach extends beyond human powers of manipulation. Scientific practices involve much more than experimental manipulations: they also include simple observation, various means of investigating past events, and many practices more distant from any form of data collection. All of these practices can be relevant to the discovery of causal patterns and to putting those patterns to use.

Another stumbling block for the idea that science aims to uncover causal patterns is that many patterns of central scientific importance are rather distant from causal *processes*. A prominent alternative to a manipulability account of causation uses the concepts of causal process and causal interaction as the basis for causal attributions. One example of this approach is Wesley Salmon's (1984) account of causation based on mark transmission. On that view, a process is causal if it is capable of carrying forward a modification, or mark. A second example is Phil Dowe's (2000) physical causation account, according to which a causal process is a world line possessing a conserved quantity. A world line is an object's trajectory through space-time, and conserved quantities are whatever our best scientific theories tell us is universally conserved (e.g., mass energy). From the perspective of a physical process account of causation, many patterns uncovered in science may seem to be non-causal. Consider again the ideal gas law, $PV = nRT$. This depicts a synchronic dependence among several features of a gas; it is not patently a depiction of causal dependence. Consistent with my description of physical process accounts of causation, Salmon for one judges the ideal gas law to be non-causal. Similarly, equilibrium models, such as optimality models and game-theoretic models in evolutionary biology, which depict structural features that together determine an equilibrium point, have also been construed as non-causal (Sober 1983; Rice 2015).

Yet there are important similarities between the dependences captured by the ideal gas law and equilibrium models and other, more obviously causal dependences. These depictions of dependence patterns are all of scientific significance in virtue of the insight and control they furnish humans. Awareness of these patterns enables recognition of how certain shifts would result in other changes in a system and, for systems that we are in the position to intervene on, the ability to accomplish changes we take to be desirable. This insight is front and center for a manipulability approach to causation. Accordingly, on that approach, the ideal gas law and equilibrium models are also taken to capture causal relationships. In a variety of circumstances, an intervention on the volume of a container, for example, would change the pressure of the gas inside according to the relationship expressed by the ideal gas law. That causal relationship is invariant over those conditions. In turn, intervening on the factors determining an equilibrium point disrupts the expected equilibrium by changing the value of the equilibrium point or eliminating it entirely. For example, Goss-Custard (1977) develops an optimality model to account for the eating habits of the redshank sandpiper (*Tringa totanus*), a bird that feeds on worms in mudflats. This sandpiper exhibits a preference for eating large worms over small worms. The model demonstrates that, if large worms and small worms are both readily available, a redshank's energy intake is maximized when large worms are chosen. This selective advantage would lead to the evolution of the preference in question. But if, for example, large worms had been historically more difficult to find in the bird's environment as it evolved (a possible intervention), the preference—or at least the degree of preference—would be different.

The reason physical process accounts of causation may not identify patterns such as these as causal is that the patterns are rather distant from actual causal history, that is, from causal processes. Equilibrium models are similar to the ideal gas law in that they do not indicate what changes in a system actually occurred in a given instance. Instead, they depict patterns in what we might call *structural causes*. A structural cause is some variable on which a phenomenon counterfactually depends in the way required by Woodward's manipulability account but that did not change to bring about the phenomenon's occurrence.[7] As we saw just above, intervening on the types of worm available by making large worms relatively unavailable would lead the redshank sandpiper to have a different evolved behavior, but change in worm availability need not, and likely did not, precede the evolution of the sandpipers' preference for large worms. Similarly, the ideal gas law shows how a change in a gas's pressure would change its temperature, even for a gas that's been maintained at a constant pressure. Notice also that the ideal gas law depicts a symmetrical dependence among variables, which might make a causal interpretation especially counterintuitive. It doesn't just indicate how a gas's temperature changes due to pressure but also how pressure changes due to temperature. But, symmetrical dependence is not problematic for structural causes in the way it presumably would be for causal processes. Symmetrical causal dependence exists whenever two variables are such that intervention on either one would change the value of the other.

The importance of what I call structural causes follows from science's emphasis on patterns. Patterns by their nature involve generalization across multiple instances. Sometimes specific causal processes recur in a patterned way. But other times, there are general patterns across instances where specific causal processes vary. The latter is exemplified by the ideal gas law and equilibrium models. By not identifying structural causes, physical process accounts of causation fail to identify these patterns. As a result, they also obscure the relationship between causal dependence and human powers of manipulation. Both of these shortcomings are significant when the focus is how science is shaped by its status as a human enterprise. Note, however, that I have not compared the merits of the manipulability account and physical process accounts as rival approaches to the metaphysics of causation. I briefly address the relationship between this discussion and the metaphysics of causation just below. But any who are distracted by my adherence to a manipulability approach to causation are free to simply rename what I call causal patterns "manipulability patterns" and set the matter of causation aside.

To summarize, our scientific representations tend to depict causal patterns instantiated by a limited range of systems to varying degrees and with exceptions.[8] I need to refine this conception of causal patterns in one more way before moving on. I said above that science depicts or otherwise capitalizes on causal patterns, but so far I have only discussed scientific products that depict patterns. It will emerge in the following two chapters that science has a wide variety of aims. For this reason, scientific projects make use of causal patterns in different ways. Many of these uses are naturally construed as depicting or representing patterns. This is the class of products I mentioned above, including apparent laws that are better understood as restricted generalizations, generalizations that explicitly employ ceteris paribus clauses, and a number of other kinds of scientific representations. But other scientific products may resist such an interpretation. The exploration of causal influence can instead involve careful causal diagnosis in one or a few specific phenomena of interest. Other projects are even more distant from representing causal patterns. For instance, the literature on mechanisms shows that other kinds of dependence, such as compositional or organizational dependence, can also be significant. And purely predictive models may have little to do with the representation of causal patterns. Yet despite all this variety, all capitalize in significant ways on causal patterns. This is because causal patterns are key to human insight into and influence over our world.

Causation is, of course, a metaphysically charged topic. There is a question of what (if anything) the causal relation really is. If one embraces a manipulability approach to causation, there are several positions one might have on the metaphysical status of that approach. One might take manipulability to be a guide to causal metaphysics; this is how Strevens (2007, 2008a) interprets Woodward's view. Alternatively, one might hold a physical relationship of causation, such as outlined by Salmon (1984) or Dowe (2000), to be metaphysically basic and conceive of facts about manipulability and difference-making as dependent on facts about fundamental causal relationships. Alyssa Ney (2009) endorses this latter position, which she calls "causal foundationalism." If I am right that a manipulability approach identifies structural causes that a physical process approach does not, a causal foundationalism would have to be supplemented with counterfactuals about fundamental causal relationships. A third option is that one might hold neither of these accounts to be metaphysically basic. Cartwright (2004, 2007), for instance, advocates causal pluralism. On that view, a variety of criteria are relevant to causal judgments, and how those criteria should be weighed depends on the particularities of individual cases.

Woodward (2003, 2008) himself is adamant that his manipulability approach is not designed to furnish a causal metaphysics. He does express doubt about the idea that causal claims are grounded in any fundamental physical causal relationships. But instead, in his view, "macroscopic causal claims (like 'chances' in a deterministic world) reflect complicated truths about an (i) underlying microphysical reality and (ii) the relationship of macroscopic agents and objects to this world" (2007, 102). This suggests that, on his view, there *is* no metaphysics of causation, on the assumption that metaphysical truths are not relative to human agents. I find this idea compelling. Nonetheless, in this book, I remain neutral regarding causal metaphysics. As far as the present project is concerned, any of these positions on the metaphysical significance of a manipulability account of causation may be correct. I signaled this above by suggesting that those with allegiances to other accounts of causation might translate my talk of "causal patterns" into "manipulability patterns." The neutrality I maintain regarding the metaphysics of causation exemplifies a position about the relationship between science and metaphysics that will be made more explicit in later chapters.

Distancing a manipulability approach to causation from causal metaphysics obviates a potentially unsettling metaphysical commitment. I have argued that the causal patterns uncovered in science hold only approximately and in limited circumstances. Allowing for the possibility of distance between these causal patterns and metaphysically basic causal relationships enables one to avoid committing to a world in which metaphysically basic causal relationships hold only approximately and in limited circumstances. Our world may indeed be perfectly law governed, these laws may or may not be causal in nature, and they may or may not govern all features of every phenomenon. But if there are such laws, science is by and large not in the business of uncovering them—not even fundamental physics. Whatever the ultimate nature of causation, the notion of causation employed in scientific reasoning is one applicable to everyday human experience, including especially how we in fact and in principle exert influence on the world. This is what results in a science useful to and comprehendible by limited human agents.

There is a different regard in which manipulability relations absolutely must be basic. Facts about manipulability are *epistemically* basic, at least relative to physical causation. These facts ground our science, providing the basis for the discovery and use of causal patterns. This is reflected in Woodward likening his concept of an intervention to an ideal experimental manipulation. It is the in-principle possibility of such manipulations that provides the foundation for our causal reasoning. This is also what makes causal dependence significant for humans. That manipulability relations are epistemically basic has two facets. This is a claim about temporal ordering: facts about manipulability are how we come to identify causal relationships. But this is also a claim about epistemic justification: we are warranted in positing causal relationships at least partly on the basis of facts about manipulability. In contrast, any successful account of physical causation will be the product of our best theories of fundamental physics. These theories, no matter how epistemically secure, require successful scientific reasoning to be already in place. This is apparent in Ney's (2009) discussion of causal foundationalism, for she gives the proviso that "to the extent that today's fundamental physics is true, it provides us with facts about causal relations that obtain at our world" (746). This epistemic limitation of any physical account of causation is significant, for as Ney also points out, today's fundamental physics is unlikely to be true, in light of the fact that it is inconsistent. And, in any case, our best fundamental physical theories are scientific products like all the rest. Human scientists developed those theories using scientific practices that capitalize on manipulability relations. Any epistemic security these theories might possess is grounded in their epistemic basis in scientific reasoning, including causal reasoning based on facts about manipulability.[9]

2.1.2 CAUSAL COMPLEXITY

One of the two starting points of this project identified in Chapter 1 is the recognition of pervasive complexity. The above argument for the centrality of the search for causal patterns in science has provided the groundwork for a more careful characterization of the relevant sense of complexity. Phenomena that are the target of scientific investigations almost always result from a wide range of diverse causal influences that interact in complicated ways. Consider the range of causal influences that come to mind for, say, the trajectory of a forest fire, animals' traits, and climate change, to name just a few examples. Even laboratory-bound phenomena are causally influenced not just by the variables targeted in the investigation but also by the controls implemented to highlight the targeted variables, as well as many more influences occurring earlier in time. I call this "causal complexity." Causal complexity may not be overtly controversial, but it has been underemphasized in philosophy, and its implications are accordingly underappreciated.

Let's look more carefully at the example of causal influences on animals' traits. This example has some themes in common with the earlier discussion of genomics projects, where I briefly touched on the broad range of influences on human traits (summarized in Table 1.1). Any evolved trait—say, the feather color(s) of some species of bird—is influenced by numerous types of causes. Feather color can result from pigmentation, how light is refracted, or a combination of these. Pigmentation and light refraction are, in turn, each subject to a number of causal influences. Genetic and epigenetic factors together result in the heritability of these traits; this is why offspring tend to have similar coloring to their parents. Numerous developmental factors also influence the expression of these traits, including the factors influencing the development of feathers themselves. Feather color has also been subject to population-level causal influences such as natural selection and drift. It is also subject to within-lifespan influences, including direct environmental influences. For instance, carotenoids, one of the main types of pigments influencing the color of feathers, are acquired through birds' diets. Some of the causal influences I have listed occur simultaneously, the occurrence of others partially overlap, and still others take place at wholly different points in the trait's causal history. Furthermore, many of these causal influences interact with one another. Genetic and developmental factors, for example, can only exert their influence in combination with one another. And parrots' green plumage results from feathers that reflect blue light and that contain yellow pigment.

Untangling such causal complexity is made more difficult by the limitations of our representations. Depicting the full gamut of causal influences in a single representation is impossible. First of all, causal histories stretch indefinitely far back in time. Even in a given period of time, there is a wide range of influences, as well as interactions among those influences. This is especially clear once structural causes are taken into account. Recall that structural causes are variables upon which a phenomenon depends but that did not change to bring about the phenomenon. Within one generation of a single population of birds, feather color is influenced by genes (probably many, separately or in combination), developmental processes, environmental factors, and influences like predation. At least some of these influences exert effects on one another: genes influence development, as does the environment; predation may influence distribution of feather colors, but feather color also influences survival. Any representation can capture only a subset of the causal influences on a phenomenon. Notice also that a delimited set of causal influences can be represented in multiple, incompatible ways. In other words, the causal space can be parsed in different ways. Genes can be represented in all their molecular glory, represented more abstractly as Mendelian units of inheritance, or more abstractly still as simply trait heritability. These representational choices additionally limit the content of any single representation.

COMPLEX CAUSALITY AND SIMPLIFIED REPRESENTATION 37

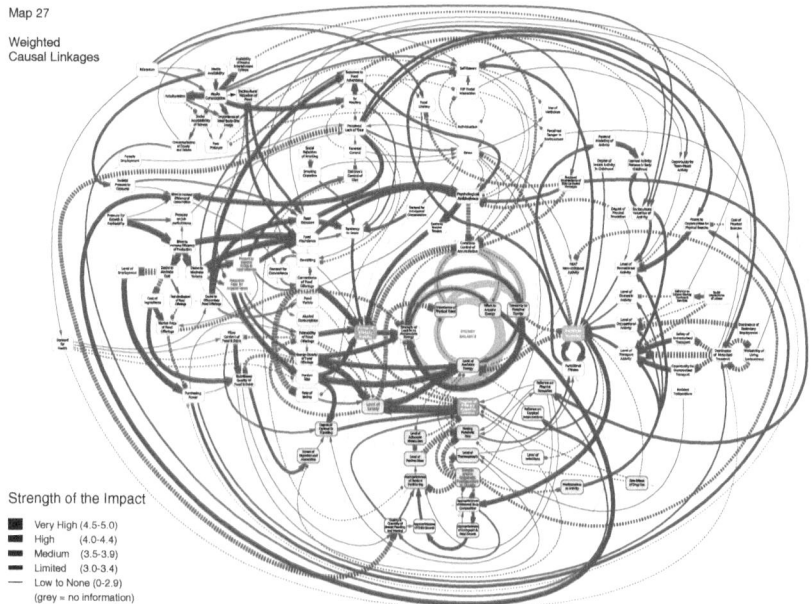

FIGURE 2.2. This map from the Obesity System Atlas shows all causal influences identified for human obesity (Vandenbroeck et al. 2007). Solid lines represent positive causal influences and dotted lines negative causal influences. Thickness of line represents the strength of the influence (whether positive or negative). Robert Skipper recommended this example. Public-sector information licensed under the Open Government License v3.0.

Consider, as an example both of causal complexity and the investigative and representational quagmires it produces, the Foresight Tackling Obesities project (Vandenbroeck et al. 2007). One of the main products of that project is the Obesity System Atlas. This atlas attempts to represent in a single "map" all of the primary causal influences on human obesity. Many of those influences are feedback loops, that is, variables that exert reciprocal influence on one another. A map including all the studied variables and the causal influences among them is shown in Figure 2.2. Figure 2.3 displays the central portion of this map at a greater resolution, since the full map is too large to show in sufficient detail here. As communicated in the introduction to the atlas, the project aims are to confirm the "inescapably systemic and messy nature" of the phenomenon of obesity, to help in the identification of relevant variables, and to help determine where in the system to intervene effectively. This emphasis on intervention accords nicely with a manipulability approach to causation, as does the focus on human action. Notice also that the causal diagram is meant to represent the general phenomenon of human obesity. It is thus best interpreted as representing a causal pattern, or perhaps a number of causal

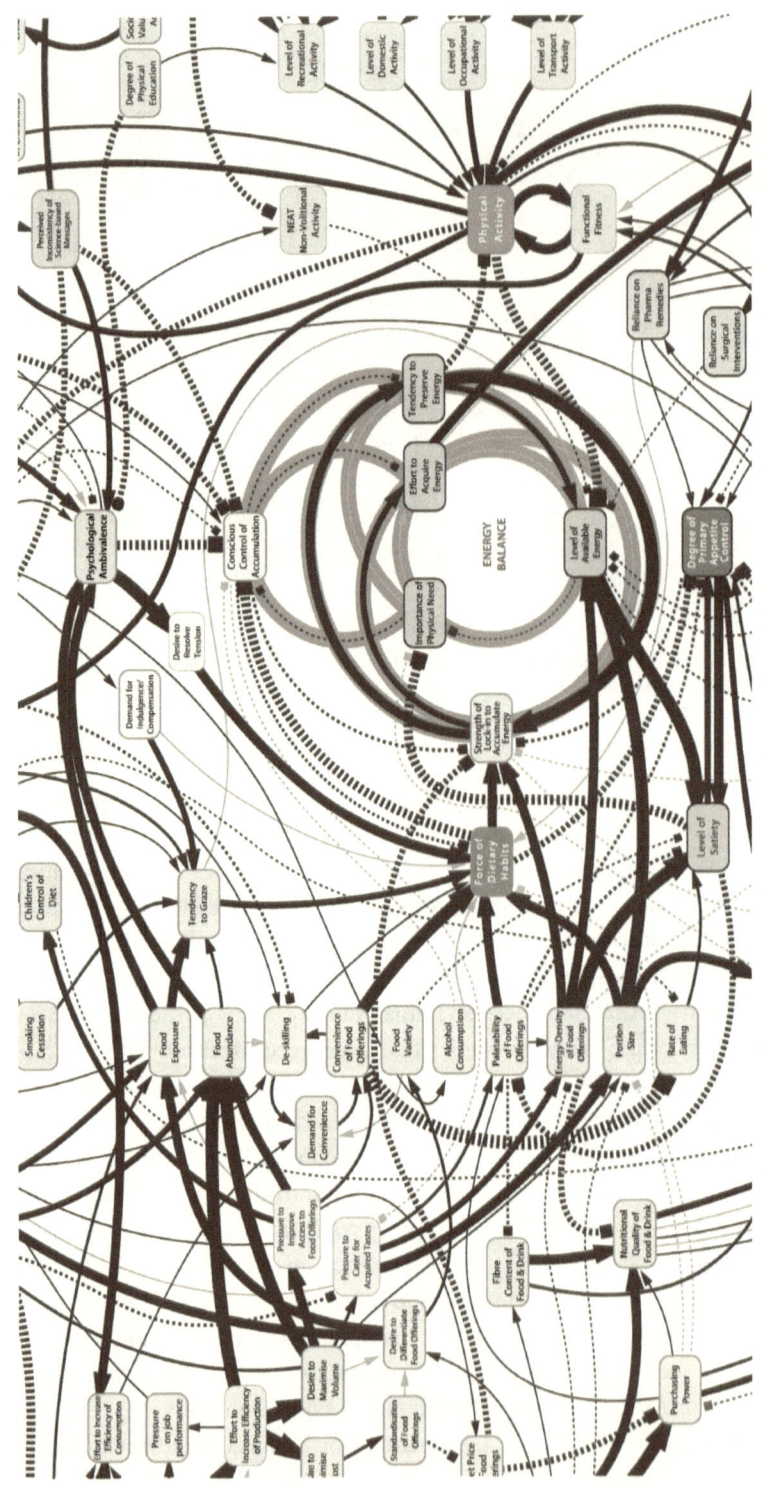

FIGURE 2.3. This is an expanded image of a central part of the Obesity System Atlas map shown in Figure 2.2. As in the primary image, solid lines represent positive causal influences and dotted lines negative causal influences. Thickness of line represents the strength of the influence (whether positive or negative). Public-sector information licensed under the Open Government License V3.0.

patterns. Many of these causes are likely to be structural causes: variables upon which the incidence of obesity depends but that did not change to precipitate some or any instances of obesity.

Despite the project's overt focus on complexity, the map shown in Figures 2.2 and 2.3 still represents only limited features of the causal influences on human obesity. Consider the following three limitations. First, this map only treats causal influences quite close to the phenomenon of obesity. Each of these factors are, in turn, subject to a number of other causal influences that are not represented in the map. Second, the causal diagram represents causal patterns, from which we should expect some degree of variation in any individual instance of obesity. Third, this map neglects significant features of each of the focal causal relationships, including its degree of invariance. The concept of invariance was introduced above in the discussion of Woodward's manipulability account of causation; this is the range of conditions over which a specific causal relationship obtains. Without information about the invariance of these causal relationships, we cannot say whether a shift in the value of some variable—depicted or not—will alter or eliminate the role of any of these variables in the causal system of obesity. For example, this map represents degree of physical education as positively influencing level of recreational activity. However, it is likely that there is some threshold of physical education that, if exceeded, does not result in further increases to recreational activity. This illustrates how the scope of invariance is just as significant to effective action as are manipulation relationships. To return to my general point, these limitations of the obesity system map do not undermine the value of that representation. Instead, this case illustrates how causal complexity far outstrips representational resources, even for depictions with the express aim of capturing the complexity.

The Obesity System Atlas also includes a number of other maps that depict different elements or layers. These provide additional insight into some of the difficulties in investigating and representing diverse causal influences. One series of maps divides the identified causal influences into several "thematic clusters," such as social psychology, individual psychology, and physiology. Influences in these different clusters are largely investigated in different fields of science. This is another dimension of causal complexity: types of causal influences on a phenomenon tend to not fall neatly into the focus of one or even a few fields of science. Another series of maps details the causal connections that exist among influences that are in different thematic clusters. These causal connections that transcend field boundaries can pose special difficulties, as fields of science tend to differ in their methodology, that is, their tools of investigation and representation. Finally, a series of maps explores which causal networks and feedback loops potential interventions would affect. This conforms to the project's stated goal of effective intervention on the system. But effective intervention requires additional investigation into, at least, the invariance of the causal relationships in question.

The scope of invariance for a causal relationship is just one of many complications in how causes exert their effects. I have also mentioned a few types of causal interdependences, including feedback loops, causal influences on some effect also influencing one another, and causes that require the presence of other factors in order to exert their influence. Sandra Mitchell (2012b) argues that, for such reasons, causal influence often fails to be modular—in other words, causal contributions are in general not additive and may thus be inseparable. She appeals to the example of a gene's causal contribution to the production of a phenotypic trait. One gene in a network may causally contribute to the production of some phenotypic trait under normal conditions but, when a different gene is disrupted, have a very different effect. This makes employing anything like Woodward's "ideal experimental manipulation" on the latter gene very difficult, if not impossible. An intervention requires that the values of all variables not on the causal path in question be held fixed, but in this case, attempting to intervene on one gene leads to changes in the causal contributions of other genes as well.[10] This does not interfere with in-principle intervention, so does not undermine causal attribution, but it does show how causal complexity renders causal analysis difficult and perhaps in some instances is even impossible.

So far I have illustrated causal complexity, and I have indicated some of its dimensions and its impacts on representation. Yet one might wonder how common causal complexity, especially in its more extreme forms, really is. I can offer two types of justification for a belief in widespread causal complexity. First, what we know about and are increasingly learning about the world corroborates this view. Complex systems approaches are applied in an increasingly wide range of scientific fields, such as developmental biology, neuroscience, fluid dynamics, and meteorology, to name just a few. Furthermore, in many fields of science, there is an ever-broadening conception of important causal factors. This is exemplified by the trajectory described in Chapter 1 from the Human Genome Project to the 1000 Genomes Project to the Human Microbiome Project, as well as by the intricacies of the Obesity System Atlas. Any number of other examples of causal complexity can be generated with the following exercise: choose any phenomenon investigated in science, then consider the types of causal influences on that phenomenon. Remember to include background conditions, causes at earlier and later time periods, and causal influences on the causes you have already identified.

When your list grows long, begin to consider how those influences overlap and influence one another. This is causal complexity. Causal complexity also is apparent, of course, in sensitivity to initial conditions, nonlinearity, and other properties of causal relationships that might be less immediately available to one's imagination.

My second justification for a belief in widespread causal complexity is the scientific significance of causal patterns—in particular, of limited patterns with exceptions. The lawlessness of science, that is, science's continuing failure to identify universal relationships of dependence, is itself a corroboration of causal complexity. With multitudinous interacting causal influences, universal, exceptionless laws of nature are either nonexistent or unidentifiable by limited human intellects. We are left with the search for causal patterns as described above. There are other views that could account for the lawlessness of science. Cartwright (1983, 1999) supposes that phenomena in our world are simply unruly. She suggests that perhaps nature is "constrained by some specific laws and by a handful of general principles, but it is not determined in detail, even statistically" (Cartwright 1983, 49). The downside to such a position is that it cannot account for the scientific successes we do find, namely, the causal patterns that have been discovered for a broad swath of phenomena of interest to us humans. It is thus the combination of lawlessness and, yet, widespread success with discovering causal patterns that to my mind suggests a world rife with complex and variable causal relationships.

2.2   Simplification by Idealization

So far in this chapter, I have made the case that science is profitably understood as a search for causal patterns in the face of causal complexity. In this section, I explore a widespread strategy for accommodating this situation, namely, with representations that idealize away much of the complexity. I begin with a discussion of how causal complexity motivates idealization. I then defend a stronger view of the significance of idealizations in science. I argue there are many intertwined reasons to idealize, and these reasons include researcher's interests as well as features of the world. I make the case that idealizations play a positive representational role, which is what distinguishes them from abstractions and other differences between a representation and the systems it's used to represent. Finally, I use this account of idealization to motivate the centrality of idealization to science. In my view, idealizations are both rampant and unchecked in science. By *rampant*, I mean that idealizations are found throughout our best scientific products, and they stand in for even crucial causal influences. By *unchecked*, I mean that little effort is put toward eliminating or even controlling these idealizations.

2.2.1 REASONS TO IDEALIZE

Phenomena are causally complex, as characterized above, and our scientific products aim to represent (or otherwise capitalize upon) causal patterns embodied by these complex phenomena. Phenomena embody many patterns, perhaps infinitely many. Indeed, this is a feature of causal complexity. The more causal influences there are, the more variables there are that produce patterned effects—either separately, in combination, or both. Further, like the ideal gas law and van der Waals equation, or bar code and bar code with 25% noise, many patterns have more exact or more general cousins. In light of all these patterns, scientists have a choice not only about which phenomena they focus on but also about which of the causal patterns embodied by those phenomena they focus on. Some phenomena are of especially broad, or particularly keen, interest for one reason or another. Scientists may collectively focus on many of the patterns such especially interesting phenomena embody. One example of this are the influences on human traits or organisms' traits more generally. As described above, researchers investigate many causal patterns in the propagation of traits, including genetic, evolutionary, developmental, environmental, and so on. But even in such cases, scientists will not be sufficiently interested in *all* the patterns, of all causal influences and in more general and more exact versions, to investigate them. Probably no one cares about the role of the availability of oxygen in bringing about some specific human trait or about the exact interplay of all causes leading to variability in the length of earthworms.

In light of the motivations I discussed earlier in this chapter for science's focus on patterns, one can make some generalizations about the kinds of patterns that tend to be of interest. Causal patterns of interest to scientists tend to be somewhat simple and somewhat general, they tend to regard causes either of intrinsic interest or of interest due to their potential for intervention, and they tend to be patterns with the right degree of specificity to provide the basis for intervention. I'm sure there are exceptions to each of these generalizations. But the scientific value of simplicity, generality, and the potential for intervention and control are well appreciated. These features are at the heart of the scientific value of causal patterns. And discerning patterns with these features requires ignoring a lot of complicating factors, that is, a lot of complexity.

Enter idealization. *Idealizations* are assumptions made without regard for whether they are true and often with full knowledge they are false. Many people are familiar with the common assumption in physics of frictionless planes and with the common assumption in economics that humans are perfectly rational agents. These are both idealizations: every plane has friction, and no human is a perfectly rational actor. We saw above that Newton's law of universal gravitation also idealizes when applied to massive bodies, for it assumes that each massive body occupies only a single point. Idealizations like these, and countless others, are found throughout science. Idealizations provide a way to set aside complicating factors to discern a causal pattern of interest. For example, the ideal gas law assumes that a gas is composed of noninteracting point particles that collide with perfect elasticity. These idealizations enable the law to ignore the (usually small) causal role of molecular size and intermolecular attraction. Idealized representation of phenomena is a technique sort of like squinting at the dot matrix images embodying bar code in Figure 2.1, so that the deviations disappear and you see only bar code.

Idealization has been receiving increasing philosophical attention, most often in the context of scientific modeling. Scientific models are explicitly intended to represent phenomena only partially. Models and the target systems they are used to represent bear some features in common, while other features of the systems are neglected or falsified in the models. This is accomplished via the use of abstractions and idealizations in characterizing the model, in order to render it simpler or easier to deal with in some other way. As William Wimsatt (1987, 2007) says, "Any model implicitly or explicitly makes simplifications, ignores variables, and simplifies or ignores interactions among the variables in the models and among possibly relevant variables not included in the model" (96).[11] These are all idealizations. Although scientific models absolutely require idealizations, idealizations are not unique to model-based science. As we have already seen, traditional "laws," such as the law of universal gravitation and the ideal gas law, are actually idealized representations. So are other scientific products, including theories, ceteris paribus laws, mechanism sketches, and so on.[12] Because of the scientific focus on causal patterns in a causally complex world, idealization is a quite general feature of science.

An initial puzzle about idealizations is the question of why, when the aim is to investigate one or more systems, one would intentionally introduce an assumption that is false of those systems. Put another way, how is introducing inaccuracies a promising first step to accurate claims about phenomena and the patterns they embody? It turns out that there are several answers to that question. Many different motivations have been suggested for the incorporation of idealizations. To mention a few prominent examples, Cartwright (1983) claims that idealizations make for more illuminating, explanatory models. Wimsatt (1981) discusses how the evaluation of several models with different idealizations can lead to the discovery of "robust" results that do not rely on any particular false assumption. Ernan McMullin (1985) discusses how idealizations can facilitate mathematical or computational tractability. Robert Batterman (2002b) emphasizes idealizations' contribution to accounting for "stable phenomenologies," or repeated general behavior. My view about idealization's role in representing causal patterns bears some obvious similarity to this last idea.

Michael Weisberg (2007, 2013) assimilates several of these views about the nature of idealization to characterize what he takes to be three distinct kinds of idealization. The first of these is Galilean idealization. This category includes simplifications merely introduced to secure the computational tractability of a model. Galilean idealizations are, accordingly, to be eliminated—and the model "de-idealized" (McMullin 1985)—if and when it proves possible to do so. The second is minimalist idealization, which is the strategy of introducing idealizations to eliminate all but the most significant causal influences on a phenomenon. The third kind of idealization Weisberg identifies is multiple-models idealization, in which several distinct, simplified models are employed to together shed light on a phenomenon. Weisberg appeals to different "representational ideals," or what sort of fidelity of representation researchers desire, to distinguish among these three types of idealization. In his view, Galilean idealizations are employed to the end of complete representation, which accounts for why they are eliminated whenever possible; minimalist idealizations facilitate the representation of crucial causes, and—significant for my purposes—multiple-models idealizations can facilitate a range of different representational ideals. Yasha Rohwer and Collin Rice (2013) argue that there is also a fourth type of idealization, overlooked by Weisberg, which they call hypothetical pattern idealization. On their view, this occurs in models that aim to represent no actual systems but instead to show what is possible and thereby contribute to theory development.

These accounts indicate an important feature of idealization in science, namely, that there are a variety of motivations behind the incorporation of idealizations into scientific representations. Weisberg's view that no single motivation accounts for all idealizations must be right. Idealizations serve a wide range of purposes in representations, and the uses of scientific representations vary greatly as well. These ideas will be additionally corroborated by the considerations and case studies introduced in this chapter and the next. Rohwer and Rice's point that Weisberg's taxonomy is not adequate to capture all of the purposes to which idealizations are put must also be right. Weisberg only allows for idealizations as temporary expedients, or to facilitate the representation of the core causal factors, or when employed in combination with other models that incorporate different idealizations. As Rohwer and Rice note, this overlooks at least one important circumstance in which idealizations are found: idealized models that obviously neglect important causal factors and yet are not employed in combination with other models. This circumstance of idealization is very common and, it will emerge, crucial to an accurate interpretation of the role of idealization in science. This is because, in my view, this style of idealized model is often used to represent the causal patterns embodied by causally complex systems.

There is, however, an important shortcoming of both Weisberg's and Rohwer and Rice's attempts to delimit kinds of idealizations. As a first step toward motivating the problem, notice that almost all of the circumstances in which idealizations occur fall within Weisberg's category of multiple-models idealization. On Weisberg's (2013) analysis, multiple-models idealization occurs when "our cognitive limitations, the complexity of the world, and constraints imposed by logic, mathematics, and the nature of representation, conspire against simultaneously achieving all of our scientific desiderata" (104). But I have argued that the effects of causal complexity and our cognitive limitations on science are quite general. Causal complexity makes it exceedingly difficult or impossible for a single representation to depict all of the significant causal influences, as is required for minimalist idealization. Our cognitive limitations, together with the world's complexity, make idealizations much more valuable than just as temporary expedients for our calculations, as is expected for Galilean idealization. So, of Weisberg's three categories, the circumstances he describes for multiple-models idealization apply to most all instances of idealization, while the others apply to few or any cases. And yet, multiple-models idealization is also the one kind of idealization that on Weisberg's view is not defined by a single representational ideal, and representational ideals were supposed to demarcate the different kinds of idealization. The circumstances of idealization that are by far the most common are, then, also the circumstances in which representational aims cannot be used to specify the usefulness of idealizations as Weisberg has suggested.

Perhaps instead, Weisberg expects that what is distinctive about multiple-models idealization is simply that multiple models are in use. But there are two different ways of interpreting this requirement: multiple models might be employed within a single research program or across the scientific enterprise as a whole. The former is a narrow sense of employing multiple models. Employing multiple models with conflicting assumptions within a single research program is a distinctive approach to science that facilitates comparisons among models' assumptions and findings, often used as the basis for robustness analysis (Wimsatt 1981; Weisberg 2006). In contrast, the latter is a broad sense of employing multiple models. The use of multiple models with conflicting assumptions in a variety of research programs across the scientific enterprise is quite common, as different research programs often focus on different aspects of phenomena, that is, different causal patterns. A research focus on one or another causal pattern will lead to a distinctive set of idealizations to facilitate the representational goal immediately at hand. It is unclear whether Weisberg intends the narrow or broad sense of employing multiple models. On one hand, he appeals to the U.S. National Weather Service's practice of employing multiple incompatible models, which is consistent with the narrow reading. On the other hand, he also cites with approval Levins's view that "communities of scientists" construct multiple models that "collectively can satisfy our scientific needs" (104), which suggests the broad reading.

This ambiguity is relevant to interpreting Rohwer and Rice's proposed amendment to Weisberg's taxonomy. Rohwer and Rice (2013) focus on idealized models that neglect significant causal influences (so do not qualify as minimalist idealizations) and that are used singly, not in combination with models incorporating conflicting assumptions (so do not seem to qualify as multiple-models idealizations). If Weisberg intends the narrow reading of the use of multiple models, then this style of idealization is not accommodated by his taxonomy. If instead he intends the wide reading, then this style of idealization counts as an instance of multiple-models idealization. But if so, this serves as an illustration of how the category of multiple-models idealization is so broad that it is a sort of dustbin category, uninformative about the features of idealizations that fall into it.

I suspect that the narrow reading of employing multiple models is the right interpretation of Weisberg's view. If so, that would leave the style of idealization upon which Rohwer and Rice focus unaccounted for in Weisberg's taxonomy. But simply introducing a fourth category of idealization, as Rohwer and Rice urge, is not the right approach. Weisberg's and Rohwer and Rice's approaches to articulating the nature of idealization are both flawed in two fundamental ways. First, they overlook how common it is for idealizations to stand in for significant causal influences and to be employed for more important and enduring reasons than temporary computational expedience. Second, both anticipate discrete categories of idealizations when there are instead quite many reasons to idealize, reasons that often occur in combination to motivate the use of any given idealization.

Consider first my claim that idealizations regularly stand in for significant causal influences and yet are employed for principled and enduring reasons. This practice pervades science. It is much more common than the use of multiple models with competing assumptions, which was Weisberg's expectation, and much more common than the investigation of hypothetical patterns, which was Rohwer and Rice's expectation. This is due to idealizations' usefulness in representing causal patterns. I suggested above that idealizations are used to set aside complicating factors to help scientists discern the causal patterns they are primarily interested in. The neglected complicating factors might well be significant causal influences in their own right, insignificant only to the focal pattern. Interest in only one or a few causal patterns a phenomenon embodies justifies ignoring other important aspects of the phenomenon to focus on what is essential to the pattern(s) of interest. Idealizations can aid in the search for causal patterns for many different reasons, as indicated just below. But some of those reasons warrant idealizations that permanently stand in for significant causal influences, simply because those causal influences are not important to researchers' main interests. Idealizations thus reflect not only causal facts but also our interests, that is, what we care about getting right. In light of science's focus on causal patterns and our causally complex world, this style of idealization is exceedingly common.

Now consider my claim that there are many reasons to idealize. In my view, a taxonomy of kinds of idealizations presents as defined categories what are better understood as overlapping motivations for idealizing. The kinds of idealization Weisberg identifies are supposed to be motivated in turn by computational tractability, the relative unimportance of some causal influences, and trade-offs in achieving scientific desiderata due to cognitive limits, complexity, and constraints of logic, mathematics, and representation (see Levins 1966). But surely these reasons to idealize are often present in combination with one another. I have argued that causal complexity and cognitive limits are quite general features of science; so too is an interest in expedience, including computational tractability. There are also additional ways in which idealizations can be expedient, for example, in enabling researchers to take modeling techniques in which they happen to already be well trained and reapply them to unrelated phenomena.

All of these reasons to idealize and more recur in various combinations throughout the scientific enterprise. Accordingly, instead of discrete kinds of idealization motivated by distinct representational aims, there are many *intertwined reasons to idealize*.[13] These reasons to idealize occur in combination in service of a variety of aims, generally related to representing or capitalizing on a causal pattern of interest. I address how idealizations positively contribute to science's aims in Chapter 4, and I suggest there that the list of

TABLE 2.1. Some of the many intertwined reasons to idealize. Together, these result in the centrality of idealizations throughout science.

	Due primarily to the world	Due primarily to scientists' features
Temporary	Exceedingly complex causal structure	Limits of computational power
	Technique happens to get traction	Familiar technique
	"Handshake" between models	Preparatory for different approach
Permanent	Computational limits	Cognitive limits
	Captures core causal influences	Limited research focus
	Enables general application	Pedagogical value

reasons to idealize is open-ended. Nonetheless, it is possible to shed some additional light on idealization by categorizing these intertwined reasons to idealize. First, some of them justify the incorporation of an idealization merely temporarily, and others justify permanent idealization. This decides whether de-idealization will ever be warranted, a difference that is also reflected in Weisberg's taxonomy. Second, some of these reasons are due primarily to features of our complex world, and others are due primarily to features of scientists themselves or of their audience. These distinctions are independent, giving rise to a categorization like what is depicted in Table 2.1.

This take on idealization accommodates the variety of justifications philosophers have suggested for idealized representation in science. Table 2.1 reflects McMullin's idea that idealizations facilitate computational tractability in two possible reasons to idealize: a temporary justification due to the current limitations of scientists' computational power, as well as a permanent justification due to absolute computational limits. The latter is akin to Batterman's (2002a) emphasis on exactly solvable models. Batterman also argues that, in some cases, general application to a number of systems is only possible with the help of idealizations. Idealizations can also be used in place of factors that don't make a difference to the phenomenon, thereby advertising their lack of importance, which is Strevens's (2008b) view; this is a permanent reason to idealize due primarily to features of the world—in particular, due to facts about causal significance. Weisberg (2013) focuses on how a specific, limited research focus motivates idealization in model-based science (for what he calls multiple-models idealization). As I have already suggested, I expect that this reason to idealize is present in a much broader range of circumstances. This is similar to the pedagogical value of simplified representation that idealizations facilitate, which Giere (1988) has noted, among others. Eric Winsberg (2010) discusses how some idealizations are introduced to facilitate combining models with different frameworks; following his terminology, this is reflected in the table as idealizations that serve as "handshakes" between models. Idealizations can also be justified, at least in a temporary use, in virtue of mundane reasons such as a researcher's previous training in a modeling technique that she then attempts to apply to unrelated phenomena. Something like this occurred when game theory was exported from rational choice theory in economics to population biology. Such mundane reasons to idealize are reflected in the table as when a technique happens to be familiar to a researcher and when a technique happens to get traction with a phenomenon of interest.

These various justifications for idealization in science provide multiple, overlapping reasons to idealize in any given instance. Consider, as an example, evolutionary game theory models, one variety of equilibrium model. One of the many idealizations common in evolutionary game theory is the assumption that a population of organisms is infinite in size. This idealization enables the neglect of genetic drift, that is, changes in frequency due to random sampling. When a game theory model is applied to a very large population, this idealization does not interfere with the accurate representation of the core causal influences, for in that case, random sampling is not a significant influence. Population size is still a structural cause, but the population is large enough that, for the purposes of genetic drift, it may as well be infinite. But capturing the core causal influences is not the sole justification for this idealization. The assumption of an infinite population also makes the game theory model tractable, so it is valuable in virtue of limits in computational power, and it reflects a research focus on natural selection and long-term evolution instead of genetic drift or short-term evolution in a few generations. There is also a conservative benefit: this idealization helps enable a model of biological evolution correspond to a model of rational choice, the origin of game theory models. When an evolutionary game theory model is applied to a smaller population, the idealization of an infinite population *does* interfere with the representation of a significant causal influence. In this case, genetic drift is a significant causal influence. But the other reasons I have described for this idealization apply to this use as well. Assuming the population is infinite still makes the model tractable, still enables the reapplication of a rational choice model to biology, and, crucially, still aids in the representation of the causal pattern of interest: the influence of natural selection on the long-term evolution of phenotypic traits.

The similarities and the differences between these two instances of the idealization of an infinite population are both notable. It is thus the individual reasons to idealize themselves, not the instances of idealization, that should be grouped as the same or different. The roles these reasons play can also vary across instances. The reasons for a given idealization can differ across instances, even very similar instances, as the example of the infinite population idealization illustrates. Additionally, an idealization might initially be introduced for one reason but retained for different reasons entirely. The idealizations required for evolutionary game theory initially enabled the reapplication of a technique familiar from the social sciences to biology, but those assumptions are only maintained because game theory found traction in representing causal patterns in natural selection's role in frequency-dependent evolution, and this is a significant research focus in population biology. In some instances, one of the reasons to idealize might be insufficient in itself to motivate an idealization, but alongside other reasons, its benefit is still felt. For instance, mere tractability is of little use if it furthers no scientific aims. Finally, an assumption might be initially adopted with the expectation of its accuracy, but when it is later discovered to be inaccurate, it is maintained as an idealization for one or more of these reasons to idealize. I expect this last scenario is quite common in the history of science. Commonsense ideas about matter and force assumed in Newtonian mechanics were once accepted as true but are now recognized to be simplifying idealizations, maintained because of the success of Newtonian mechanics in dealing with relatively large objects moving at relatively slow speeds.

2.2.2 IDEALIZATIONS' REPRESENTATIONAL ROLE

I now shift my attention from the motivations for idealizing to the role idealizations play in models and other scientific representations. The consensus in the literature on model-based science is that idealized models can represent despite their false assumptions. Models are explicitly designed to be simpler than the phenomena they represent, and those simplifications are accomplished at least in part with the incorporation of idealizations. To formulate this as a general point about representation, scientific representations can incorporate idealizations, or false assumptions, and yet still succeed at their representational aims. Here I advocate a stronger view. I argue that idealizations themselves play a positive representational role. Idealized models and other scientific products do not represent despite their idealizations but partly in virtue of those idealizations. Moreover, idealizations aid in representation not simply by what they eliminate, such as noise or non-central influences, but in virtue of what they add, that is, their positive representational content.[14]

Notice that not every difference between a representation and the phenomena it represents counts as an idealization. Take as an example of a simple, nonscientific representation the official map of the Bay Area Rapid Transit

FIGURE 2.4. On maps of the BART system, the stops within San Francisco are equidistant and in a straight line. On the official BART map, it's also the case that the Richmond-Millbrae line is colored red. The former is an idealization; the latter is not. Credit: San Francisco Bay Area Rapid Transit District/www.bart.gov/CC BY 3.0.

(BART) system in the San Francisco Bay Area. A monochromatic version of this map appears in Figure 2.4. Consider two features of the BART map. First, in the original map, the Richmond-Millbrae line is colored red. Second, all stops within San Francisco are equidistant and in a straight line. Both are differences between the map and the BART system, but only the second counts as an idealization. One might articulate the relevant difference between these inaccurate features of the map by pointing out that the map represents the BART lines as if their stops within San Francisco were equidistant and in a straight line. In contrast, the map does not represent the Richmond line as if it were red. This same way of distinguishing between idealizations and other differences between representations and represented phenomena exists in science. Newton's law of universal gravitation represents massive bodies as if they are point masses when they are not; this is a difference between the law and and actual systems that qualifies as an idealization. Newton's law can also be characterized as a simple mathematical formalism, whereas no system to which it applies can be. This is a difference between the law and the system, but it is not an idealization.

Idealizations are distinguished from other differences between a representation and what is represented in virtue of representing a system as if it were some way that it is not. I propose, then, that idealizations actually play a positive representational role. An idealization does represent targeted phenomena, for it represents those phenomena as if they were some way they are not. I call this *representation as-if*. This idea fits comfortably with the paradigmatic type of idealization, namely, representing a system as if it were ideal in some regard. A surface may be represented as if it were a frictionless plane, an individual as if she were a perfectly rational agent, or a population of organisms as if it were infinite in size. Not all idealizations are naturally construed as representing a system as if it were ideal; it would be strange to call a point mass an ideal physical body. The general idea here is that an idealization represents a system as if it possessed one or more features that it does not. For this reason, idealizations must be accorded a positive representational role.

As I mentioned above, Strevens (2008b) affords a role to idealizations in representation, but that role is only in virtue of what idealizations eliminate. He says, "All idealizations, I suggest, work in the same way: an idealization does not assert, as it appears to, that some nonfactual factor is relevant...rather, it asserts that some actual factor is irrelevant" (316). On Strevens's view, for example, the assumption that a population is infinite is (always and only) a way of communicating that population size is irrelevant. I disagree, of course, that idealizations always and only play this role. As I argued in the previous section, idealizations are present for a variety of reasons, and they frequently stand in for significant causal influences. And, to the present point, in my view, idealizations have positive representational content. They don't aid in representation simply by eliminating irrelevant factors but in virtue of what they *do* represent. In direct contradiction to the quote from Strevens, idealizations communicate positive relevance, not merely irrelevance. An idealization asserts the nature of a factor's relevance by saying something false about that factor—by representing it as if it were some way it isn't. In doing so, idealizations are fulfilling a subtle, but positive, representational purpose.

In brief, idealizations' positive representational role is to indicate the nature of a factor's relevance to the focal causal pattern. How this is accomplished with representation as-if can be identified by briefly considering philosophical accounts of representation in general. In this book, I remain neutral on the nature of the representation relation, but considering both of the two main camps on that issue is enlightening here. One camp, similarity-based approaches, takes the basis of representation to be commonalities between the representation and that which is represented. The other camp takes representation to be grounded in intentional or pragmatic considerations. Both kinds of approaches can accommodate, and help make sense of, representation as-if.

For similarity-based approaches, the representation relation exists in virtue of the commonality between representations and what they represent. The required commonality might be taken to be isomorphism (van Fraassen 1980), analogy (Hesse 1966), or similarity (Giere 1988; Weisberg 2013). It may initially be puzzling how an idealization—by definition dissimilar from a system—can have a representational relationship to that system. But successful idealizations do bear certain similarities to the systems they help to represent. Idealizations represent phenomena as if they had features they don't, but those misrepresentations are useful insofar as there are *functional similarities*—similarities in causal role or behavior—between the idealization and features of the represented phenomena. This idea can be motivated especially well by considering the example of effective population size, or N_e, which is a common variable in population genetics. A popular population genetics textbook (Hartl and Clark 1997) defines effective population size as "the number of individuals in a theoretically ideal population having the same magnitude of random genetic drift as the actual population" (121). N_e is thus an idealization. It is a property attributed to a fictional entity, a "theoretically ideal" population, but used to characterize a real entity. Effective population size represents a real population of organisms as if it were a size that it is not. This is useful because, despite the dissimilarities between the effective and actual population sizes, they have a functional similarity. A real population of some actual size N experiences genetic drift identically to an ideal population of size N_e. The actual and effective population sizes are different numbers, and the actual and ideal populations may be different in any number of other regards, but this particular behavioral result is the same. If similarity is at the heart of representation, then such a functional similarity enables representation as-if.

In contrast to accounts of representation based solely on similarity or isomorphism, some philosophers have developed accounts that ground the representation relationship on intentional or pragmatic considerations. This kind of an approach to representation can also accommodate the representational role of idealizations.[15] On this approach, an idealization stands in a representational relationship to a system simply in virtue of a researcher's decision to use it in this way. This idea would undoubtedly be cashed out in different ways by different specific accounts of representation. Representation as-if may derive from the researcher's intentions, that is, from an analogy that is drawn between an idealization and some feature of real systems or from pragmatic considerations of the idealization's usefulness in this regard.

Accounts of representation seem to increasingly incorporate some form of both elements, similarity and intentions, to ground the representation relation. Newer versions of the isomorphism- and similarity-based accounts developed by van Fraassen (2008) and Giere (2004), respectively, also incorporate considerations of intention. I suspect that both elements are also involved in representation as-if. Broadening the range of permissible types of similarity in the way I have suggested generates the need to specify which functional similarities are important, and this depends on researchers' intentions. In light of its particular functional similarity to actual population size, effective population size is useful in representing causal patterns in the role of drift. It's utterly unhelpful for representation in many other research contexts. On the other hand, intentions and pragmatic considerations seem to be in themselves insufficient—suitable similarities must exist for any representational relationship, including representation as-if. This is *why* effective population size isn't helpful in many research contexts but can be helpful in representing causal patterns involving drift.

I have suggested that idealizations' positive representational role is to indicate the nature of a factor's relevance to the focal causal pattern. This discussion of the requirements for representation as-if helps make sense of that idea. Returning once more to the example of effective population size, this represents a population as if it were a size it is not, which is justified by the functional similarity between how the actual population and an ideal population of that size experience genetic drift. This is useful when researchers want to capture causal patterns involving drift. The idealization of effective population size captures the combined relevance to genetic drift of population size, population structure, skewed sex ratios, and a host of other features of the population while otherwise enabling the neglect of those features. Many of these neglected features are causally important to evolutionary outcomes, but they are not important to patterns in how genetic drift influences evolutionary outcomes. This idealization is thus warranted only insofar as patterns in drift's influence are focal (to the exclusion of other patterns). Similarly, the BART map pictured in Figure 2.4 represents the stops in San Francisco as if they were equidistant and in a straight line because BART travelers' circumstances ordinarily lead them to focus solely on the spatial and temporal order of stops, and this idealization and the system's actual properties are similar in the regards that are relevant to this ordering. In science, an idealization represents a feature of a phenomenon as if it were other than it is to demonstrate the feature's exact relevance to a focal causal pattern.

Recall from the discussion of patterns earlier in this chapter that a system might embody a pattern imperfectly. Additionally, many otherwise important features of a system may not be relevant to its embodiment of a given pattern. Causal complexity renders both of these scenarios commonplace. Any phenomenon results from myriad causal influences, and some of those influences may interfere with the action of the focal causal influence. Idealization, representation as-if, enables depicting a system as the embodiment of a causal pattern, which involves setting aside features of the system that are incidental to the causal pattern and that may actually make that embodiment imperfect. This is the utility, or at least one utility, of representing a phenomenon as if it were some way that in fact it is not.

Many distinguish idealizations from abstractions, where the former is misconstrual of a system's features and the latter is neglect of a system's features. As Cartwright (1989) says, in regard to the idealization that a plane is frictionless, "The model may leave out some features altogether which do not matter to the motion, like the color of the ball. But it must say something, albeit something idealizing, about all the factors which are relevant" (187). Martin Jones (2005) works out this view more fully. My view about idealizations' positive representational role provides a compatible distinction between the practices of abstraction and idealization. With idealization, the target system is represented as if it had some features it does not, whereas with abstraction, some features of the system are straightforwardly omitted, that is, they are ignored without consequence for the representation. It is important to distinguish this sense of omission—ignoring without representational consequences—from omission in the sense of failing to explicitly reference. It is the former sense that is definitive of abstractions; many idealizations are also omissions in the latter sense. Evolutionary game theory models hardly ever specify that the population is infinite in size. Instead, this idealization is generally left implicit. Nonetheless, it is a requisite assumption whenever genetic drift is not taken into account. Any time drift is neglected, the population is represented as if it were infinite. As we have seen, in a causally complex world, far-flung causal relevance—and ensuing distortion—is commonplace. This makes idealization widespread and renders its representational role quite significant. This also makes it so that, time and again, intended abstractions have been discovered to be covert idealizations.

Because idealizations represent things as if they were different than they are, they are in a sense fictions. Idealizations positively represent phenomena, and they do so falsely. To idealize is in this sense to fictionalize.[16] Accordingly, the scientific importance of representation as-if results in the elevation of fictional properties and entities—things like infinite populations and frictionless planes—to central roles in scientific investigations. In physics and chemistry, for example, there are central references to ideal gases, defined as theoretical gases composed of noninteracting point particles in random motion, and perfect gases, which are theoretical gases additionally simplified with other stipulations, including that they are not chemically reactive. Another example of the central role of such constructs is provided by the idealization of effective population size. Consider how Lande and Barrowclough (1987), in a heavily cited chapter in a collection on conservation and population biology, employ the concept. These biologists say that the purpose of their chapter is to "show how the effective size of a population, the pattern of natural selection, and rates of mutation interact to determine the amount and kinds of genetic variation maintained" (87). This indicates that they are investigating, in part, the causal influence of effective population size—an idealization, the property of a fictional, ideal population.

The term "fiction" has been used in various connections to idealization, so I should clarify how the present idea relates to a few other references to fictions. First, some, including Godfrey-Smith (2006), have developed the idea that models should be taken to be fictions in the sense of imaginary concrete systems, on par with literary fictions. This is a different issue entirely. My claim is that idealizations themselves, wherever they are found in science, qualify as fictions, and this does not entail a view about the ontological status of models. One may grant that idealizations are fictions and yet reject the idea that models incorporating idealizations describe fictional worlds. The latter idea, though, would offer one way to account for some fictional properties and entities introduced by idealizations taking on a scientific life of their own, as I have suggested they do. Second, in contrast to what I have said, Winsberg (2010) distinguishes idealizations from fictions. In his view, idealizations represent inexactly, whereas fictions represent falsely. The idea here is that some assumptions get things more or less right, whereas other assumptions are radically incorrect. Beyond mere terminology, the question here is the similarity or dissimilarity of these types of assumptions. I have suggested that the very same idealizations are employed in both kinds of circumstances. An infinite population size might be posited when the population is so large it might as well be infinite, or it might be posited when the population is smaller but the research focus is a causal pattern to which genetic drift does not contribute. Winsberg would call the first an idealization and the second a fiction, but I think both uses have the same central features. Regardless of the flagrancy of the falsification, an idealization is only warranted when it accommodates the obtaining research interests. Third, Alisa Bokulich (2009) develops a view of how fictions, taken to be idealized models, can explain. Her view also emphasizes the scientific centrality of fictions, and her conclusions are complementary to my claim of idealization's positive representational role. She says, "Some fictions can capture in their fictional representation real patterns of structural dependencies in the world" (107). Notice that Bokulich's focus on fictions also emphasizes their importance to capturing patterns of dependence, which I take to be causal patterns.

I have argued that idealizations play a positive representational role, despite their falsity. Idealized models and other scientific products do not represent despite their idealizations but partly because of those idealizations. And idealizations aid in representation not simply by eliminating irrelevant factors but in virtue of their positive representational content. An idealization represents a feature of a phenomenon as if it were other than it is, to the end of capturing the feature's precise relevance to the focal causal pattern (but, importantly, not to the phenomenon itself). Philosophical accounts of scientific representation have the resources to naturally accommodate this positive representational role of idealizations. Idealizations, and the fictional entities and properties they posit, are integral to scientific practice in virtue of representation as-if.

2.2.3 RAMPANT AND UNCHECKED IDEALIZATION

In this chapter, I have argued that science is in the business of discovering causal patterns in the face of causal complexity and that idealizations play a particularly valuable role in enabling the representation of causal patterns. I have suggested that there are many different reasons to idealize that, in various combinations, motivate idealization. And I have argued that idealizations themselves play a positive representational role, representing systems as if they had features they do not. Together, these points about idealization indicate the reasons why idealizations are absolutely central to science, and they afford a strong, positive role to this feature of science. Idealization is both rampant and unchecked in science.

Consider, first, the idea that idealizations are *rampant* in science. By this I mean that idealizations exist throughout our best scientific representations, and they stand in even for important causal influences. This is because causal complexity renders idealizations incredibly useful to the pursuit and representation of causal patterns. Science is tailored to human needs and thus human limitations, which leads to a focus on rather simple patterns that contribute to human understanding and influence. Representing a single causal pattern embodied by a complex phenomenon requires some degree of misrepresentation of the phenomenon, which facilitates the neglect of features not central to the pattern of immediate interest. So idealizations are pervasive. But some features not central to a focal causal pattern are important causal influences in their own right. Those influences are idealized in just the same way. Idealization thus does not signal causal insignificance, simply insignificance to the present research focus.

The second part of my view about the centrality of idealizations is that these rampant idealizations are also *unchecked*. By this I mean that there is little focus on eliminating idealizations or even on controlling their influence. Consider first the claim that there is little focus on eliminating idealizations, or on what has been called de-idealization. This idea is not particularly controversial among philosophers who emphasize idealization's role in science. Several such philosophers, including Wimsatt (1987), Batterman (2002a), Bokulich (2009) and Weisberg (2007, 2013), hold that at least some idealizations are permanent, in the sense that reducing or removing the idealizations to increase the accuracy of representations is not taken to be a goal. I agree, of course. Several of the intertwined reasons to idealize identified above are due to permanent features of human scientists and their limitations or permanent features of the complex world we inhabit. Some idealizations can, for example, best facilitate the representation of core causal influences or can enable a focus on causes of primary research interest. Both of these are ways to eliminate extraneous features to represent a causal pattern. Idealizations can also be of permanent value in light of absolute computational limits or our human cognitive limits. When idealizations are justified in part by one or more permanent considerations like these, which is exceedingly common due to the world's complexity and humans' limitations, eliminating them would not constitute scientific progress.

Idealizations are also unchecked in the sense that there is little focus on controlling their influence; this idea is stronger and more controversial. Philosophical justifications for permanent idealizations generally fall into two types. One might think that idealizations are warranted in virtue of features of the world and in particular facts about causal significance and insignificance. This is Strevens's (2008b) view and describes what Weisberg calls minimalist idealization. For both, idealizations are supposed to stand in only for insignificant features of a system, or non-difference-makers. Accordingly, this does not accommodate the persistent idealization of significant causal influences. Batterman (2002a) endorses a similar view, claiming that idealizations must be justified by showing that the details they get wrong are safely neglected. Here too idealized factors must be unimportant to render idealizations unproblematic, that is, to keep them in check. An alternative justification sometimes offered for permanent idealization depends instead on scientific practices to keep idealizations in check. Thus, as we have seen, Weisberg's multiple-models idealization is permitted to neglect significant factors but only when alternative models, employing alternative idealizations, are also developed. Rohwer and Rice (2013) make the case that other permanent idealizations are permissible but only in the quite limited representational role of representing merely possible systems. Both of these are proposals for how to keep idealizations in check with scientific practices. In contrast to both of these ways for keeping idealizations in check, my view sanctions idealization even of central causal influences, on a permanent basis, and without taking any steps to hold in check the resulting misrepresentation. Indeed, I have argued that this misrepresentation—representation as-if—positively contributes to the representation of actual systems. This does not require the compensatory use of other, more veridical representations.

Many philosophers agree that idealizations are widespread in science. The controversy surrounds the circumstances in which those idealizations are legitimately employed and the degree to which it is preferable to eliminate them. I have developed a very strong view of idealization in this chapter. This view legitimates the permanent use of idealizations in many roles, including a central role in representing actual phenomena, even when they stand in for significant causes and without measures taken to control their influence. The extensiveness of idealization on this view exceeds what many philosophers expect. Weisberg's view accommodates continuing idealization only when the idealizations do not stand in for central causal influences or else when multiple models are employed, with other models representing the idealized factors. In contrast, I have suggested that significant causal influences are regularly idealized without recourse to multiple models; indeed, I suspect this is by far the most common circumstance of idealization. This is the case for highly idealized models that are not applied to any particular systems, as Rohwer and Rice (2013) emphasize, but also for idealized representations of causal patterns embodied by particular systems, predictive models that sacrifice realistic causal representation, and other uses still. There is another regard in which the centrality of idealizations exceeds what many philosophers anticipate. Fictional properties and entities arising from idealizations, such as effective population size and ideal and perfect gases, can take on a scientific life of their own. As I briefly illustrated with the example of effective population size, research can be devoted to exploring the roles of these constructs in much the same manner as their real analogs.

However, rampant and unchecked idealization does not mean unprincipled idealization. Idealizations reflect researchers' interests, and they serve those interests in the face of specifiable cognitive, computational, and other limitations. Intertwined reasons to idealize, like those identified above, thus motivate the inclusion of idealizations and determine their nature. An idealization would be pointless without motivation by one or more such reasons. Worse, an idealization could be undermining by impeding a primary aim. Additionally, because idealizations play a positive representational role by representing as-if, the nature of that representation must be appropriate for the focal causal pattern; to other causal details of the phenomenon, even if they are of little interest; and to the aims and method of the research. An infinite population is not a helpful idealization when the research focus is genetic drift, for it can't aid in representing patterns in the role of drift. Nor is it a helpful idealization when the research focus is natural selection, but a recent bottleneck in population size swamps all other evolutionary influences; population size isn't a research focus, but the bottleneck obliterates any pattern in the role of natural selection. Nor is it a helpful idealization in the logistic model of population growth, which requires reference to the actual (finite) population size. An idealization that is inappropriate in any of these regards will never be made in the first place or will be eliminated whenever possible. The same is true if the research focus or method changes in a way that warrants a different approach to idealization.

Idealizations are absolutely central to science, in light of the search for causal patterns in the face of causal complexity and human limitations. Idealizations stand in even for important causal influences, when setting those influences aside helps in the identification of a causal pattern of interest. Representing causal patterns requires stage setting. Causal patterns hold only within a limited range of circumstances and have deviations and exceptions even within that range. Phenomena must be represented in ways that reflect the causal pattern and that maximize its salience to human researchers. This requires making idealizing assumptions, thereby neglecting other important features of phenomena. These circumstances give rise to the many intertwined reasons to idealize, both temporary and permanent, some due primarily to features of scientists and others primarily to features of the world. When an idealization is present merely for temporary reasons, there may be a scientific benefit to de-idealization when those reasons no longer obtain. But this is uncommon. Most idealizations are employed in part for permanent reasons, including especially that they help scientists identify causal patterns focal to their research. These idealizations are of permanent scientific value; their elimination would not improve our science. Moreover, they are not shortcomings of a representation. Instead, idealization has a range of benefits. Idealizations allow researchers to focus myopically on exactly what they care about in a phenomenon. They can be tailored to an audience, which is why their role in textbooks has long been appreciated. They enable the reapplication of existing approaches to disparate phenomena. And, different idealizations can be adopted to fit different projects—the identification of broader or more nuanced patterns, patterns featuring different causes singly or in combination, the prediction of outcomes of complex processes we only dimly understand, and so on.

3

The Diversity of Scientific Projects

A fairly unified picture of scientific practice emerged in Chapter 2. There I argued that depicting and otherwise capitalizing on causal patterns is a central feature of science. More specifically, scientific research quite often involves a search for causal patterns in the face of causal complexity. These patterns are limited because they only hold in some circumstances and because phenomena deviate from the pattern to some degree—sometimes to such a great degree that they are exceptions to the pattern. It is possible for phenomena to embody multiple, distinct causal patterns, and in our causally complex world, most or all phenomena in fact do. In part due to this situation, scientific projects involve continuing idealization; idealization is both rampant and unchecked. Idealization, which I have argued plays a positive role via representation as-if, facilitates a focus on one causal pattern at the expense of representing other features of phenomena, including other causal patterns.

However, this unified description conceals the full diversity of science's projects. Attending to even a few contemporary scientific research programs reveals great variation in aims and methods. In this chapter, I consider scientific investigations of cooperation in behavioral ecology; a variety of approaches to investigating human aggression; and, more briefly, fluid dynamics, quantum physics, and climate science. This survey demonstrates that any generalizations about science must accommodate great diversity. Differences among the scientific investigations I examine include the aims of the research, the role played by data, and how the project connects to other scientific research. And yet, despite this diversity, all have something in common: they employ idealized representations. This further confirms the extent of idealization in science. Behavioral ecology models of cooperation are great illustrations of idealized representations of causal patterns, with little emphasis on any specific systems. Investigations of human aggression are, in contrast, focused on a much more specific phenomenon. Accordingly, these are helpful for assessing whether and how idealization is involved in a different style of research. Finally, I briefly consider other philosophers' work on fluid dynamics, quantum physics, and climate modeling—three types of research in the physical sciences. This enables me to further assess the role of idealization and the search for causal patterns outside the so-called special sciences. Together, the great diversity of scientific projects and the extent of idealization begin to cast doubt on the idea that science uniformly proceeds toward truth, unified representation, or any other unitary aim. Accordingly, this discussion sets up the analysis of science's aims that occupies Chapter 4.

3.1 Broad Patterns: Modeling Cooperation

One prominent research focus in behavioral ecology is the evolution of cooperative behavior, especially among animals. Historically there have been three main approaches to accounting for cooperation: a variety of group selection models, kin selection models, and models of reciprocal altruism. Perhaps the best-known research into the evolution of cooperation is modeling reciprocal altruism using the prisoner's dilemma, a game theory model that was first applied to the evolution of cooperative behavior by Axelrod and Hamilton (1981). The prisoner's dilemma represents a way in which cooperation can emerge among self-interested individuals when two different behaviors are more beneficial immediately and in the long run. In its evolutionary interpretation, "self-interest" simply means that the cooperation arises by means of maximizing each individual's fitness. This should not be understood as a claim about psychological motivation. The prisoner's dilemma model applies to encounters among organisms when one strategy immediately benefits one individual's fitness at an immediate cost to the other individual's fitness, whereas a different strategy is less immediately advantageous but maximizes both individuals' fitness in the long run, provided there are many such interactions. The latter strategy, because it is less immediately advantageous to the actor and less immediately costly to the other individual, is considered cooperative.[1]

Here is an illustration of cooperative behavior that can be modeled using the prisoner's dilemma. Vampire bats (*Desmodus rotundas*) risk dying from starvation or dehydration if they do not feed every night. When a bat returns to its roost after an unsuccessful hunt, some bat that did hunt successfully will regurgitate a portion of the blood it ingested to feed the starving bat. This enables the at-risk bat to survive to hunt another night. Wilkinson (1984)

TABLE 3.1. (a) This is a matrix representation of a vampire bat's immediate payoffs based on whether it shares its hunting spoils and whether another bat in turn shares with it. This is a classic prisoner's dilemma; it is most advantageous for a bat to refuse to share, so long as other bats nonetheless share with it. (b) This represents the long-term payoff to a bat under certain assumptions. In a total of 10 exchanges, the sharing bat is assumed (conservatively) to encounter nine sharing bats, whereas the nonsharing bat is assumed (conservatively) to encounter only one sharing bat. This is because, as Wilkinson (1984) discovered, most bats engage in reciprocal altruism. They share their hunting spoils but only with bats who are also known to share. The cumulative payoffs make clear that, in these circumstances and relevantly similar ones, it is in a bat's self-interest to share.

Focal bat	When another bat	
	Shares	Doesn't share
Shares	3	0
Doesn't share	5	1

(a) Immediate payoff, one exchange

Focal bat	When other bats	
	Share	Don't share
Shares	$(9 \times 3) + (1 \times 0) = 27$	
Doesn't share	$(1 \times 5) + (9 \times 1) = 14$	

(b) Long-term payoff, 10 exchanges

demonstrates that this arrangement is an instance of reciprocal altruism. By sacrificing some of its own food, a donor bat acts in a way that benefits the recipient at an immediate cost to itself. But this maximizes the donor's own fitness in the long run, for it ensures that other bats will in turn feed it when it is the one facing starvation. Indeed, Wilkinson gathered data corroborating reciprocation: bats preferentially feed their roost-mates, recipients of blood are in turn more likely to serve as donors, and the only non-reciprocating bat observed was never fed blood by others. Table 3.1 shows the relative payoffs of the prisoner's dilemma and how these payoffs accrue in a case of reciprocal altruism.

The prisoner's dilemma is a highly idealized model of the evolution of cooperation. A number of assumptions are required, assumptions that are false of many or all target systems. These idealizations include the assumptions that reproduction is asexual and that the population size is infinite, which together ensure a perfect relationship between a trait's success and its prevalence in the population. They also include the assumption that the fitness effects of traits are constant across individuals and circumstances, which enables the use of a simple payoff matrix. These are just a few examples of the extensive idealizations required for the prisoner's dilemma to be used as a model of evolution.[2]

Other behavioral ecology approaches to modeling the evolution of cooperation differ in their setups and thus in what they posit as the evolutionary causes of cooperative behavior. Kin selection models represent how cooperation can emerge among related individuals in virtue of beneficial effects on inclusive fitness, that is, the fitness benefits one's relatives derive from one's own cooperative actions. Group selection models, in turn, show how cooperation can evolve as a result of how it improves the fitness of a group of organisms taken as a whole. Despite these different specific focuses, though, all of these approaches have the same basic aim: to account for the role of natural selection in producing cooperative behavior. This aim determines what idealizations are useful. Accordingly, idealizations very much like those described for the prisoner's dilemma are general across all these approaches to modeling the evolution of cooperation and, indeed, behavioral ecology models targeting other kinds of traits as well. For example, the assumptions of an infinite population and asexual reproduction allow any influence of genetic drift or recombination on the evolutionary outcome to be ignored. This means that selective benefit is taken to be the only contributor to evolutionary success, or at least the only contributor that is represented. This is why all these models can, crucially, equate a trait's success with its prevalence in the population. These and other idealizations greatly simplify the representation, which enables the use of modeling techniques—including the prisoner's dilemma and evolutionary game theory in general—that would otherwise be unavailable.

It is instructive to consider the directions in which this behavioral ecology research into cooperation is progressing. There is an ever-expanding variety of heavily idealized models of circumstances that generate cooperative behavior, including other game theory models and different approaches entirely. There is also work on the relationships among these alternative models. Kerr and Godfrey-Smith (2002), for instance, demonstrate the mathematical equivalence of game theory models and group selection models. In other words, they show that group selection models' calculation of fitness effects for groups and for individuals yields results equivalent to game theory's calculation of frequency-dependent fitness effects for individuals. There is also exploration of the variability of features that classic approaches assume to be static. Worden and Levin (2007) investigate the conditions in which populations might actually evolve away from the payoff structure of the prisoner's dilemma, that is, when the fitness effects of behaviors might actually be changed. For a vivid example of what this might look like in the scenario of vampire bats sharing food, imagine that bats began immediately attacking any nonsharer. This would remove the dilemma by making food sharing directly and immediately advantageous. There are also attempts to represent interaction among evolutionary influences and within-lifespan influences on cooperation. Akçay et al. (2009) find that care for the well-being of others can evolve by individual selection, and this mutual regard can in turn lead to cooperative behavior. All of these examples of research progress are further developments and explorations of highly idealized, mathematical representations. Many regard the mathematical machinery itself, without any direct attention to the representational relationship between the model and actual cooperative behavior.

Indeed, there are few examples of research designed to more accurately represent actual instances of the evolution of cooperation. When specific populations are discussed, they are almost always simply used as exemplar cases to motivate a general, idealized model. This is the case for Wilkinson's (1984) study of food sharing in vampire bats; the purpose of that investigation was to show that the trait better fits a reciprocal altruism model than a kin selection model. Axelrod and Hamilton (1981) briefly mention a number of biological scenarios, most instances of mutualism, to indicate the types of circumstances in which the prisoner's dilemma model might apply. Exemplar cases may not even be actual phenomena but simple imaginary scenarios. Maynard Smith (1982) employs this strategy to illustrate various game structures for evolutionary game theory. When introducing the hawk-dove game, he says, "Imagine that two animals are contesting a resource ... imagine, for example, that the 'resource' is a territory in a favourable habitat" (11).

In light of this description of the research trajectory, it may come as no surprise that little work has been invested in the de-idealization of these models. De-idealization in this case would involve incorporating into a behavioral ecology model more realistic assumptions about the features of a population in which a cooperative trait of some kind has evolved, the genetics governing the inheritance of a cooperative trait, or other central causal influences that existing models wholly idealize. But the extremely unrealistic assumptions of these models are seldom replaced with more accurate assumptions. The exception to this is that some models incorporate a mode of genetic transmission in lieu of the assumption of asexual reproduction and direct trait propagation. Sometimes the rationale is provided that the inclusion of genetic details is important for the researchers' interests. Curnow and Ayres (2007) develop such a model, and they state in their introduction that "a full population genetic approach is needed if the rate of change of allele frequencies is of interest" (68). That is, if one cares not just about what trait evolves in the long term but also how quickly the genetic influences on the trait evolve, then one must represent something about the trait's genetic transmission. However, even then, when the idealization of asexual reproduction is eliminated, it is replaced with simple, arbitrary genetic dynamics. These are an idealization in their own right, for they no more reflect the nuances of actual genetic transmission than do models that simply assume asexual reproduction.

Taken together, these features of the research trajectory suggest that there is little emphasis on applying these models to accurately represent specific instances of cooperative behavior. What, then, is their purpose? Above I suggested that these models are used to account for the role of natural selection in producing cooperative behavior. I more fully develop this idea in Potochnik (2009). There I call this the weak use of these models, to be distinguished from the aim of accounting for *all* significant evolutionary influences. Natural selection is just one of many causal influences on any given evolutionary outcome. To focus on its causal role in producing a trait invariably involves neglecting other causes. These behavioral ecology models are thus naturally viewed as attempts to capture causal patterns. As indicated by the limited attention to the specifics of any one evolutionary outcome, their aim is to depict patterns in how natural selection can, in general, causally contribute to the emergence of cooperation.

Because these models of cooperation ignore so many other causal influences in order to focus on the role of natural selection, they have a limited range of application and limited accuracy of most any evolved trait. In Chapter 2, I suggested that representations of causal patterns generally have both these limitations. This feature of behavioral ecology models is corroborated by the fact that some research into these models aims to demonstrate the proper range of their application. Eshel and Feldman (2001), for instance, analyze the conditions in which evolution can be expected to lead to the equilibria predicted by game theory. They find that game theory models can accurately represent the dynamics of long-term evolution but not short-term evolution. I suspect that the limited accuracy of behavioral ecology models of cooperation as descriptions of any one instance of evolved cooperation is one reason there is little emphasis on application to specific evolutionary phenomena.[3] Different behavioral ecology models can be judged by whether they do a better or worse job of representing the role of natural selection in a particular instance of evolved cooperation, but even the best representation neglects or falsifies many, even most, of the significant causal dynamics. If this is right, then the relationship of this research program to empirical investigation is indirect, established by the causal patterns these models are used to represent.

Two main forms of empirical corroboration might be pursued for these models of cooperation (see Lloyd 1988), and both fit comfortably with my interpretation of these models as representations of causal patterns involving natural selection. First, researchers sometimes empirically confirm the satisfactoriness of a model's assumptions for a particular population. For example, Wilkinson (1984) corroborates some assumptions that must be satisfied in order for food sharing among vampire bats to embody the causal pattern of reciprocal altruism. One of these is the assumption that pairwise interactions occur among the same individuals for an extended period of time. Wilkinson corroborated this by evaluating vampire bats' lifespans and the nature of their social groups. Sometimes, though, this kind of corroboration does not involve demonstrating the accuracy of assumptions. As we have seen, many assumptions of these models are idealizations. In those cases, corroborating an assumption involves determining whether a population can be represented *as if* it possessed some feature it does not. An example of this variety of corroboration is when Eshel and Feldman (2001), referenced above, argue that a population facing the same environmental circumstances for a long period of time validates the idealizing assumption of simple heritability via asexual reproduction. They don't mean that these assumptions are accurate in those circumstances. In most cases, heredity is not simple, and reproduction is often sexual, so it involves genetic recombination. But populations in a constant environment for a long period of time can be represented as if traits propagated asexually. This is because, in those circumstances, complex genetic transmission and recombination can be expected to have little influence on the role of natural selection.

In the second form of empirical corroboration, researchers seek to corroborate a model's prediction. Wilkinson (1984) does this as well. He evaluates the fit of the prisoner's dilemma model to the phenomenon of food sharing in vampire bats by showing that, as the model predicts, blood is shared preferentially with bats from the same population who were in dire need, and blood is withheld from bats who previously refused to share with others. Generally what is sought in this form of corroboration is a qualitative fit, that is, that a model predicts the right form of behavior or, for quantitative traits, a behavior in roughly the right range. This is taken to demonstrate that the representation of natural selection's causal role is satisfactory, even if the outcome is also shaped in part by other, unrepresented evolutionary influences. On my interpretation of these models' representational role, satisfactory representation of a natural-selection causal pattern requires, first, that the posited influence is present—a particular form of selective influence—and, second, that unaccounted-for causal influences do not obliterate the anticipated causal pattern—that the selective influence affects the evolutionary outcome as the model anticipates. This accounts for why empirical corroboration is sought for models' assumptions and qualitative predictions. Aside from this, the bulk of the attention in this research program is focused not on empirical confirmation of specific features of models in particular cases but on exploration of the nature and extent of the focal causal pattern across phenomena.

Behavioral ecology research into the evolution of cooperation has featured in two different philosophical debates, and these debates yield additional insight into their nature. First, there is a significant literature in philosophy of biology on the various approaches to accounting for cooperative and altruistic behavior of animals. This is the main type of evolutionary phenomenon addressed in the levels of selection debate, which questions whether natural selection acts solely on genes, individuals, or a variety of entities, including also groups. Sober and Wilson (1998) are influential advocates of the group selection approach as a "unified evolutionary theory of social behavior." They argue that cooperative behavior modeled using evolutionary game theory or kin selection theory should be acknowledged to involve group selection. Kerr and Godfrey-Smith (2002), in turn, show that an individualist perspective, such as provided by evolutionary game theory, is mathematically equivalent to a group perspective. These authors argue that each approach has a distinct "heuristic advantage," so it is useful to move between these perspectives. This debate can be construed as regarding whether there is a uniquely right behavioral ecology approach to modeling the evolution of cooperative behavior, either for a specific instance of evolved behavior or across evolutionary history. Different models make salient different patterns. Some, like Sober and Wilson (1998), think that one or another model more accurately represents the real pattern of natural selection's influence. Others, like Kerr and Godfrey-Smith (2002), hold that multiple patterns are embodied and that each is valuable by illuminating different causal relationships among phenomena. Both sides of this debate can be construed as views about the causal patterns represented in these models.

Second, there is also a substantial philosophical literature on criticisms of an evolutionary biology methodology that Gould and Lewontin (1979) disparagingly dubbed "adaptationism." As applied to behavioral ecology models like those considered here, the concern is that these models simply assume that traits, such as cooperative behavior, are the direct products of natural selection. A host of other possible influences on traits, such as genetic and developmental constraints, are entirely neglected. In a sense, this criticism is accurate. I have suggested that these models do focus on the influence of natural selection to the exclusion of other causal influences on evolution. This enables the representation of causal patterns featuring natural selection, which is what I have termed the weak use of these models. But, crucially, this use in no way inhibits the representation of other causal patterns embodied by the very same phenomena. We have seen that any given system embodies many causal patterns and that some reasons to idealize are due to particular research interests. Other biologists, with different interests and perhaps different expertise, undoubtedly aim to discover other causal patterns in these same evolutionary phenomena. Just as certainly, their representations do not employ the same idealizations as the behavioral ecology models we have considered. The debate over adaptationism thus draws attention to the partiality of behavioral ecology models of cooperation.

In summary, idealizations pervade behavioral ecology models of cooperation. Little attention is paid to the accurate representation of any one system, that is, of any single cooperative trait in a particular population. The aim instead seems to be to account for the causal role(s) of natural selection in producing cooperative behaviors in general. Accordingly, the main goal is not to generate accurate predictions or to accurately represent systems but to create partial, skewed representations. The partialness and skewing furthers the depiction of patterns in the causal role of natural selection. That behavioral ecology maintains distinct approaches to modeling cooperation suggests that multiple causal patterns provide insight into the role of natural selection in the evolution of cooperative behavior in general and, perhaps, even in individual instances of evolved cooperation.

3.2 A Specific Phenomenon: Variation in Human Aggression

A reasonable response one might have to my analysis in the previous section is to blame the heavy use of idealizations on the features of behavioral ecology in particular. The focus—the evolution of cooperative behavior—is an exceedingly broad phenomenon, encompassing a great number of specific behavioral traits across a wide variety of species. Further, the modeling techniques brought to bear are in large part analytic, mathematical models. These features together result in the value of highly idealized representations of general causal patterns, and because the phenomena are so diverse, these representations may not be very accurate of any specific systems. Accordingly, let us now shift our attention to a very different type of scientific project: investigations of a specific human behavioral trait. Human behavior is investigated across several fields of biological and social science. I focus on human aggression, which receives attention across the fields that address human behavior. Longino (2013) has analyzed this research in great depth, and my discussion will partly rely on and partly respond to her analysis. This discussion will have a somewhat different character from my treatment of behavioral ecology models of cooperation, since instead of one research program, I discuss a variety of approaches to studying aggression.

The field of behavioral genetics focuses on establishing the relative contribution of genetic variation and environmental variation to the variation in some behavior observed within a population. Both inherited characteristics and environment contribute to behavioral differences among individuals; these are often glibly referred to as as nature and nurture. The basic idea of this approach is that their relative contributions to variation in a particular trait are distinguishable in cases when they vary independently. Two common research approaches in behavioral genetics are twin and adoption studies. Twin studies compare the behavioral similarities between fraternal and identical twins, since the latter but not the former are genetically identical. Adoption studies compare behavioral similarities between twins raised together and twins raised apart. These comparisons are used to generate a heritability estimate, that is, an estimate of the degree to which a trait's variability among individuals is due to genetic inheritance. Accordingly, the focus of this approach to human aggression is not on the question of what leads to instances of aggressive behavior or to the propensity to behave aggressively. The focus is instead on what contributes to the variation among levels of aggression exhibited by different individuals. This variation is the focal phenomenon for all of the approaches I consider here.

A plausible construal of the aim of behavioral genetics research is to locate where one might look for clear causal patterns, given a phenomenon that is undeniably causally complex. If variation in aggression were found to have very high heritability or very low heritability, this would direct attention respectively toward or away from genetic research into aggression. And, indeed, aggression has been found to be heritable, with genetic influences accounting for roughly 40% to 50% of variation (Craig and Halton 2009; Pavlov et al. 2012). This may in part account for why aggression is a research focus in genetics as well as in fields that investigate how the social environment influences behavior. In their introduction to a special issue on recent advances in behavioral genetics, Althoff and Hudziak (2011) say, by way of justifying the continuing significance of behavioral genetics, that "because nature and nurture are intimately entwined, untangling them is a critical and important endeavor" (5). They suggest that behavioral genetics also provides insight into the variety of interactions among genes and environment.

Molecular genetics, in turn, attempts to establish associations between specific genes and traits, and this approach is also applied to human behavioral traits. In molecular genetics research into human aggression, much attention has been directed to the role of genetic influences on serotonin regulation, such as genetic factors influencing monoamine oxidase A (MAOA), a serotonin and norepinephrine metabolizing enzyme. As surveyed by Manuck et al. (2000), this research might involve any of several different approaches. In gene knockout experiments in animals, a gene is deleted and the effects on the animal's traits observed. Knockout experiments in mice have shown that deletion of a gene influencing the production of MAOA does result in mice exhibiting heightened levels of aggression. Human families with an extreme level of some trait can be tested to establish whether they have a mutation of a gene with suspected influence on that trait. Investigation of a large family exhibiting highly aggressive behavior has implicated the same genetic influence on MAOA as the knockout experiments. Currently, the most common approach in molecular genetics research is genome-wide association studies, which seek to establish a correlation between particular variants of a gene and levels of aggression and related behavioral traits. Manuck et al. (2000) employ this approach and find a statistically significant relationship between different variants of a gene dubbed MAOA-uVNTR and level of aggression.

Neurobiological approaches to human aggression have already made an appearance in the brief description of molecular genetics research. Determining neurological factors relevant to variation in aggression levels is often an intermediary step to the search for genetic influences on aggression, as attention to MAOA and the genetic influences on it illustrates. Here too serotonin metabolism is a main focus; another main focus is the role of testosterone. Another neurobiological approach is to use functional neuroimaging to identify brain areas that are distinctive in highly aggressive individuals. Findings of this research include reduced functioning in the frontal lobe and differences in amygdala function, where calculated aggression has been linked to reduced function and impulsive aggression to heightened function (Glenn and Raine 2014).

The final approach to investigating human aggression that I consider here focuses on how specific environmental influences affect aggressive behaviors. The research considered to this point draws significantly from methodology and research findings in fields of biology, but some of the research into environmental influences bears more resemblance to the social sciences. A common strategy is to seek correlations between specific features of individuals' environments, including especially their social environments, and aggressive behavior or some proxy for it. For instance, much attention has been devoted to establishing a relationship between childhood exposure to violent television and heightened aggression in young adulthood (e.g., Huesmann et al. 2003). Glenn and Raine (2014) survey a number of prenatal and childhood environmental influences that have been linked to heightened aggression in later years, including exposure to lead, manganese, or nicotine; fetal maldevelopment; and childhood malnutrition. They also mention that some of these factors have been found to have a greater effect in combination with features of the social environment like maternal rejection and social adversity.

This overview of some of the primary approaches to investigating human aggression suggests a potential relationship among the different types of

THE DIVERSITY OF SCIENTIFIC PROJECTS

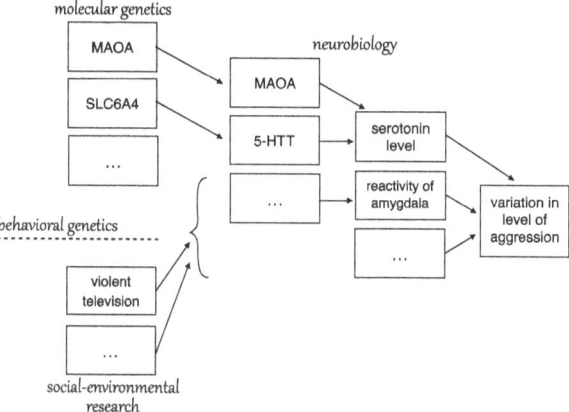

FIGURE 3.1. This is a sketch of the tidy relationship that may be posited of human aggression research. The arrows represent causal influence. The plain typeface indicates factors found to influence level of aggression, and the script indicates the areas of research that focus on those influences.

investigations, something like the following division of labor. Behavioral genetics divides causes of variation in aggressivity into environmental and genetic categories. On the genetic side, neurobiology sorts out the ways in which our nervous system influences levels of aggression, and then molecular genetics determines the genes that influence those neural features. Social science research, in turn, establishes the (nonheritable) environmental influences on aggressivity. This influence is presumably also mediated by neural factors, but identifying those factors does not have the significance for this research that it does for the molecular genetics research. This construal of the relationship among these different investigations is supported by work like Pavlov et al. (2012), which incorporates research from all these different areas into a single analysis. Pavlov et al. discuss heritability estimates of aggressivity on the basis of twin studies and connect this to the results of genome-wide association studies. The bulk of their analysis regards specific genetic influences on finely specified neurological factors. Pavlov et al. say, "These techniques promise to identify neural pathways through which these [genetic] variants contribute to the emergence of variability in behavior and disease" (74). Although nothing is said about specific environmental influences on levels of aggression, the researchers do note that neurological factors can mediate environmental influence. This anticipated relationship among investigations, drawing on Pavlov et al.'s analysis and other research discussed here, is depicted in Figure 3.1.

However, there are several difficulties with this interpretation of the relationship among these investigations. To begin with, human molecular genetics has been confronting a difficulty with missing heritability, and this undermines any tidy relationship this research might have with behavioral genetics or neurobiology. Missing heritability is when molecular biology cannot identify any specific genetic influences on a trait that behavioral genetics research has found to be heritable. Maher (2008) suggests some possible reasons for this difficulty: perhaps many rare variants play a role; perhaps epistasis, in which one gene modulates the effects of another, is common; perhaps environmental influence has been misattributed to genes; or there may even be nongenetic inherited influences on gene expression, called epigenetics. All of these possibilities complicate the expectation that behavioral genetics establishes the degree of genetic heritability and molecular genetics subsequently provides the means of inheritance. Significant epigenetic influences or enough misattributed environmental influence would undermine that expectation entirely. Craig and Halton (2009) stress this difficulty for investigations of the genetic basis of aggression in particular. They say, "The overwhelming conclusion ... is that there are few, if any, loci with large effect size, and it is becoming increasingly obvious that it will be necessary to consider the impact of genes, not in isolation, but as part of a multifactorial miasma including both other genetic factors and the environment" (108).

Two more general difficulties with the tidy division of labor shown in Figure 3.1 are also suggested by this problem of missing heritability. First, different types of causal influences may well interact, that is, mediate each other's influence, and some may even exert a causal influence on one another, creating feedback loops. For example, if there is epigenetic influence on aggression, environmental conditions influence genetic expression in a way that is then inherited by one's offspring. This undermines the division of influences into separable environmental and genetic (i.e., inherited) components. We should expect such causal interaction and feedback loops, for human aggression is undeniably a causally complex phenomenon. This causal complexity is reflected in Craig and Halton's (2009) call for considering genes as part of a "multifactorial miasma." The second difficulty is that, by focusing on a subset of causal influences, each of these approaches neglects many other types of influences. This neglect is accomplished by—you guessed it—extensive idealizations. Idealizations incorrectly represent the causal factors they stand in for. But the idealized factors are certainly causally relevant and, moreover, are likely to interact with and perhaps even directly affect the focal causal factors. These idealizations thus result in the represented causal relationships being of only limited accuracy.

Let's consider a few examples of how idealizations limit the accuracy of the causal dynamics represented in these approaches. Behavioral genetics' apportioning of variation into genetic and environmental elements requires the assumption that twins reared apart experience wholly different environments. But this is not so, for pairs of twins experience the same intrauterine environments (Maher 2008; Longino 2013). Accordingly, that approach inaccurately attributes any intrauterine environmental influences to genes. Aggression research in molecular genetics and neurobiology focuses on how genetic variation influences neural features implicated in aggressivity variation, usually neglecting entirely the certain environmental influences on the nervous system and the possibility of environmental influence on genetic expression itself. This research thus represents the causal pathway from gene to nervous system to behavior as more direct than it is and genetic factors as more significant than they are. Research into the role of social environment, in turn, treats variation in environment as if it were the sole source of variation in aggressivity. Genetic differences among individuals are implicitly assumed to vary in a way that balances them out across different social environments, but this neglects the potential causal significance of genetic variation and, even more problematically, the possibility of causal interaction between genetic and environmental factors.

As these examples illustrate, each of these approaches to research into human aggression ignores nonfocal causal influences and interactions with nonfocal influences. They accomplish this with the use of idealizations, as illustrated by the examples in the previous paragraph. This diminishes the accuracy of how each of these approaches represents the phenomenon of variation in human aggression. A variety of reasons—intertwined reasons—motivate these idealizations. Surely part of the motivation for the extensive idealizations is simply the need for simplifying assumptions to use any one of these approaches to study variation in human aggression. Perhaps this is seen as merely preparatory research, with the idea of achieving a fuller and more integrated account of aggression as the research progresses. This would be a temporary reason to idealize. Something like this might well be so for the molecular genetics and neurobiology research, at least from the perspective of molecular genetics. That approach regularly makes use of neurobiological findings to direct research into possible genetic contributors to aggression and makes use of molecular genetic findings to suggest possible neural contributions to aggression. But molecular genetics research is something of an outlier in this regard. The other approaches proceed largely in isolation from one another, without concern for the inaccuracies resulting from their simplifying idealizations. In my view, this is because they are targeting different causal patterns, which provides a permanent reason to idealize indexed to specific research interests. Each approach idealizes in order to represent how one or a few factors influence human aggression, setting aside myriad other influences to the extent possible.

In these scientific investigations of a specific phenomenon, as in the investigation of the more general phenomenon of cooperative behavior discussed above, idealizations are used to enable the representation of causal patterns. The different approaches to investigating human aggression vary in the types of causes for which patterns are sought. Each approach idealizes many significant features of this causally complex phenomenon to represent the focal causal pattern. Accordingly, no single approach captures the whole story. Nor is it the case that each approach provides one part of a consistent, complete account of human aggression, as would be the case if Figure 3.1 accurately reflected the relationship among the approaches. What is represented is instead a number of different causal patterns. Because they focus on the causal role of different factors, these patterns have little to do with one another, even though they are embodied by the same phenomenon. On this interpretation of their aim, these approaches are revealed to be by and large successful. With the exception of molecular genetics' challenge of missing heritability, most succeed at identifying patterns in the causal influences on aggressivity. But despite the success of these accounts, the actual, rich causal history of aggression levels in our human populations undoubtedly deviates to some degree from any one of these patterns. Furthermore, as we have seen, these accounts do not comfortably combine into a unified treatment of human aggression. For these two reasons, if an interpretation of the aim of these approaches in line with Figure 3.1 were adopted, the current state of this research would appear to be in rather bad shape. Interpreting this research as aiming at causal patterns thus accommodates its successes, clarifies its limitations, and provides an interpretation of its recognized shortcomings.

Longino (2013) discusses in much greater depth the disconnects and inconsistencies among these and other scientific approaches to studying human behavior, especially aggression and sexuality. She details the incompatible assumptions made in the process of employing these different approaches, and she argues that these result in incommensurable parsings of the causal influences on human behavior. On the basis of that analysis, Longino emphasizes the impossibility of "a single encompassing and unified approach" (125). She is, then, also opposed to the idea that there is a tidy relationship among these approaches like what is depicted in Figure 3.1. I agree with Longino that these approaches cannot be combined into a single unified approach. Further, I believe that my view that these approaches focus on different causal patterns, along with my argument for intertwined reasons to idealize, suggests why this is so. And yet I have one concern with Longino's conclusion that I mention now and explore in more depth later, in Chapter 7. Longino's emphasis on the limitations and disconnects among these different approaches fails to account for how these very limitations tend to spur scientific developments to *create* interconnections.

To see what I have in mind, consider that Craig and Halton (2009), who describe the difficulty of missing heritability for human aggression, emphasize that it is necessary to consider "gene by environmental interactions" to puzzle out individual genes' causal contribution to aggression. This is a way to focus directly on causal interaction among genetic and environmental influences, the omission of which is one of the significant limitations of other approaches. Longino (2013) also discusses this approach of studying genetic-environmental interactions, dubbed $G \times E$, and as she demonstrates, this approach has its own limitations. But this call for an alternative approach in light of the limitations of existing research projects illustrates an important feature of science. When existing approaches collectively fail to capture some causal interaction that is prioritized by scientists' research interests, an approach will be sought to capture that causal pattern as well. Aside from $G \times E$, a few other examples of this include evolutionary developmental biology, the simultaneous representation of evolutionary and within-lifespan influences on cooperation, and models of frequency-dependent selection pressures themselves evolving. These last two examples were mentioned in the previous section.

James Tabery (2014) develops an account of the scientific research into human behavior and, in particular, into the interaction of genes and environment that parts ways with Longino's analysis along similar lines to what I have suggested. He analyzes what he calls a deep "explanatory divide" between the two main approaches to studying the causal role of genes and environment in determining human behavior. The main difference between Tabery's and my positions is that he argues this explanatory divide should be overcome, whereas I think many scientists have good reason to stay on their own separate sides of such divides. Translated into my own terminology, my point is that even integrative approaches motivated by the shortcomings of existing research projects do not supersede their predecessors by providing a single, integrated treatment of the focal phenomena. Instead, new approaches supplement existing research by representing a causal pattern that has not yet been adequately captured. This construal of the relationship among approaches to investigating human aggression is a middle ground between the tidy view of integration and Tabery's call for overcoming divides on one hand and, on the other, Longino's claim of incommensurability. It is depicted in Figure 3.2. This figure shows five of the many causal patterns

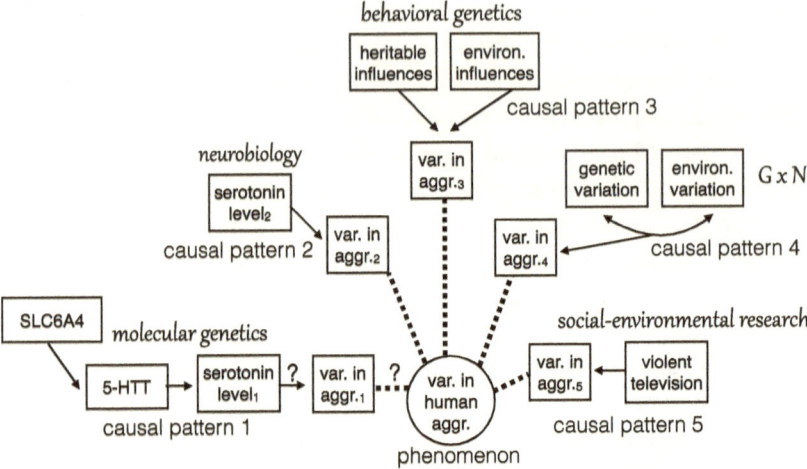

FIGURE 3.2. On my construal, the different approaches to research on human aggression fail to connect to one another. Instead, each approach posits a distinct causal pattern. The arrows here represent causal influence, the plain typeface in boxes indicates elements of causal patterns, and the script indicates the areas of research that focus on those causal patterns. The dotted lines indicate that the single phenomenon of variation in human aggression embodies these different patterns. Notice the question marks accompanying the example causal pattern posited in molecular genetics. This is because the problem of missing heritability may indicate that this causal pattern is not embodied by the phenomenon.

investigated by the different approaches to research on human aggression. Research succeeds if it shows that the pattern it investigates is embodied by the focal phenomenon. What this figure doesn't represent well is that approaches are on occasion partly unified (note that "serotonin level" appears in causal patterns investigated by both the molecular genetics and neurobiology research) and that gaps can spur new, more integrated approaches—as with the $G \times E$ approach.

So far I have argued that, despite human aggression research's focus on a single specific phenomenon, this research is still best interpreted as generating information about causal patterns. In this regard, it is similar to the behavioral ecology research on cooperative behavior surveyed in the previous section. However, an interesting contrast between these projects emerges when considering the different roles played by empirical investigation. I have suggested that empirical investigation is not central to models of the evolution of cooperation. The use of data is much more relevant to research into human aggression. All of the approaches we have surveyed employ experimentation or other data collection techniques to corroborate or disconfirm specific hypotheses about causal influences on aggressive behavior. This must be due to the goal of accounting for the specific phenomenon of human aggression, in contrast to behavioral ecology research into the evolution of cooperative behavior of any kind, in any species. Yet, because idealizing assumptions are also used in aggression research to limit the causal factors under consideration, the type of data considered by any one of these approaches is highly constrained. Data are more or less limited to determining the nature of the causal pattern in question; significantly less is done to corroborate idealized assumptions. Perhaps this is because many of the idealized assumptions in each of these investigations of human aggression are implicitly or explicitly acknowledged to distort the representation.

This use of data suggests that, despite the focus of this research on a more specific phenomenon, little or no attention is directed toward increasing the overall accuracy of the representation of causal influences. Instead, these different strands of research are pursued largely in isolation from one another, each with its own body of idealized assumptions. This accords with the relationship among the research approaches posited above, namely, the separate pursuit of different causal patterns. As we have seen, idealizations facilitate the representation of causal patterns in part by representing as inactive causal factors that are recognized to be active, some of which are even recognized to interact with the focal causal factors. Only a neglected cause's apparent interference with the pattern of primary interest or a burgeoning interest in a different, more complex pattern spurs reconsideration of idealizations. The aim of representing a pattern involving only one or a few focal causes makes data collection feasible. It also facilitates understanding of and, perhaps, influence on levels of aggression via control of the focal causes. Different causal patterns, and thus different research approaches, hold the potential for different kinds of interventions, be they neurobiological, hormonal, or social in nature.

In summary, idealizations are pervasive in these approaches to researching human aggression, despite the rather specific focus on variation in the level of aggression among individuals. The surveyed research programs limit themselves to establishing the causal role of certain types of influences. This is accomplished by seeking empirical confirmation of specific hypotheses, while extensive idealizations remain. Little attention is devoted to confirming the adequacy of these idealizations, and they often do a poor job of reflecting the causal influence of neglected factors. Few if any attempts are made to replace these idealizations with more realistic assumptions. Instead, new approaches are sometimes developed to explore the causal patterns that the idealizations of current approaches collectively overlook. There is a greater focus on accurate representation of a specific phenomenon than with behavioral ecology models of cooperation and a resulting increase in the relevance of this research to intervention. But because the focus of any given approach is limited to a subset of causal factors, the result is still partial, skewed representation.

3.3 Predictions and Idealizations in the Physical Sciences

I have now surveyed a handful of research programs in the biological and social sciences and made the case that, despite the variety of their agendas, all are productively understood as attempts to represent causal patterns, facilitated by significant idealizations. Yet it would be reasonable to attribute those characteristics to the fields of science from which my examples are drawn. Indeed, some have argued that these sorts of characteristics distinguish these fields, the so-called special sciences, from physics. Alexander Rosenberg (1994), for example, defends a view according to which physics and chemistry aim to describe reality, whereas the life sciences and social sciences merely aim to furnish us with tools for better controlling phenomena in these domains. I thus now turn my attention briefly to a few example research programs in the physical sciences, in particular, fluid dynamics, quantum physics, and climate science. Here I rely on other philosophers' work, as I have little expertise in these fields. My aim is to continue to survey some of the diversity in how scientific projects proceed and to consider how my emphasis on causal patterns and idealization fares for a few projects in the physical sciences. This will show that the research practices on which I have focused are not special to the "special sciences."

Batterman (2009) makes the case that idealizations are essential to a full grasp of some physical phenomena. One of his primary examples is the fluid dynamics account of water dripping from a faucet. When water drips, there is a hydrodynamical discontinuity that consists in the liquid separating into two or more droplets. Research into this discontinuity regards a very general but precisely defined phenomenon; it is potentially relevant to any number of instances in which a single mass of liquid spontaneously divides into multiple masses. One focus in this research is on the shape of a fluid's surface when the discontinuity occurs. This is characterized by simultaneous shifts in the velocity of the fluid and in the curvature of the fluid surface at the location of the break. In Batterman's words, this is "a difficult and complex moving boundary value problem" (434); he details several equations that together define the fluid's velocity and surface curvature.

Yet certain idealizations can be made in virtue of the system approaching a hydrodynamical discontinuity—that is, the breaking point—and those idealizations simplify matters considerably, facilitating exact solutions to the set of equations. The first idealization Batterman (2009) notes is that the water can be treated as if it were a vertical line. This is because its axial extension (length) is much greater than its radial extension (width) as the discontinuity is approached. This idealization enables a one-dimensional solution to an otherwise quite complicated set of equations (the Navier-Stokes equations). Batterman also points out that, approaching the discontinuity, the relationship among surface tension, viscous forces, and inertial forces is such that acceleration due to gravity can be wholly neglected, even though it is present. Finally, Batterman stresses the significance of the observation that the axial and radial lengths both become, at the discontinuity, arbitrarily small. For this reason, the shape of the fluid at the point of discontinuity is expected to not depend on the size of the system, that is, of the fluid mass and surface area. Batterman emphasizes how these idealizations can be justified by properties of the phenomenon in question. According to him, these idealizations serve a methodological role, for they result in an exactly solvable equation, as well as an explanatory role, for they demonstrate why different fluids, forming droplets in a variety of different circumstances, have the same shape at the breaking point.

This research is similar to behavioral ecology research into cooperation insofar as both attempt a highly general account of a broad phenomenon. In the process, both incorporate idealizations that enable the neglect of many influences on the phenomenon. Batterman (2009) shows how a set of complex equations can be simplified by the removal of parameters that represent influences on the phenomenon. As in behavioral ecology, there is some attention toward justifying these idealizations. However, the considerations Batterman surveys that enable the simplifying assumptions are of a different sort than in the behavioral ecology case. They regard geometrical relationships, such as between the axial and radial extension of a water drop, and claims about causal significance motivated by considering the models themselves—that is, the structural equations—such as the claim that surface tension, viscous forces, and inertial forces are of equal importance. In contrast, we saw that idealizations in models of cooperation are often justified by direct empirical corroboration or else left unjustified. This difference may arise from a better grasp of the relevant contributing factors in fluid dynamics research, in part because there is greater uniformity in those factors and their contributions across systems.

This research project is also naturally construed as providing a generalization about a causal pattern. The pattern regards how a fluid's shape as it breaks into droplets is determined by its viscosity. It seems limitations and exceptions to this pattern are expected as well. When Batterman (2009) indicates how the result can be expected to generalize, he specifies that "to a large extent and for a wide range of fluids, this turns out to be the case" (435). This example from fluid dynamics is, then, similar in many ways to the behavioral ecology case, with the difference that this phenomenon permits a more unitary, and more general, treatment. This difference, and Batterman's focus on cases like these, might account for his emphasis on the permanent use of idealizations only to stand in for non-central causes. Other cases, including the behavioral ecology research and human aggression research described above, as well as the climate science research described below, employ idealizations to stand in for important causes in their own right, when those causes are not central to the research focus.

Next let's consider a research project closer to fundamental microphysics. Bokulich (2008, 2009) examines the continuing significance of classical mechanics even now that, as a theory of fundamental physics, it has been superseded by quantum mechanics. Quantum physics is a paradigmatic physical science, in the sense that it involves laboratory research, regards the tiniest bits and pieces that compose everything, and thus is taken to be wholly general. In contrast to the example research discussed earlier in this chapter, this research does proceed, at least sometimes, in the way traditionally emphasized in philosophy. Namely, predictions are derived from a general theory and then tested to corroborate or disconfirm the theory. And yet, Bokulich argues that there is a deep divide in at least some instances between general theory and the explanation of phenomena. Even as quantum mechanics is the reigning theory, having superseded classical mechanics, one sees the ineliminable use of classical mechanics in explaining some quantum phenomena.

For example, following the account of Bokulich (2009), the electron orbits posited in classical mechanics, and even the idea that electrons follow definite trajectories of any kind, are rejected in quantum physics. An electron is instead taken to be a cloud of probability density around the nucleus of an atom. But the latter cannot account for a peculiar effect. Typically, an atom absorbs photons in a particular pattern until the atom is ionized and can absorb no more photons. But when an atom is also subject to a strong magnetic field, it continues to absorb photons after the ionization threshold has been passed. This, along with the resulting absorption pattern, isn't expected from a quantum perspective. Its eventual explanation, closed orbit theory, uses classical mechanics to calculate the possible electron orbits of the ionized atom experiencing the magnetic field. These possible orbits are then further limited according to the interference patterns produced by the electrons, now interpreted as waves. The remaining orbits account for the absorption pattern beyond the ionization threshold and, it turns out, a number of related phenomena. This explanation relies crucially on the classical account of electrons as particles following fixed orbits, as well as the quantum account of electrons as waves exhibiting interference patterns.

As Bokulich (2009) emphasizes, the appeal to classical electrons following fixed orbits is an idealization, a fiction. (I briefly described Bokulich's view of idealizations and fictions in Chapter 2.) She stresses that this fiction is not merely a calculation tool but provides "genuine insight into the way the world is" (103). I would phrase this as the representation of electrons as if they were particles following classical orbits. Recall that representation as-if occurs when a particular research focus limits the significance of some feature of the system such that an idealization can productively represent it falsely. This is in virtue of a relevant functional similarity between the real feature and the idealization. The resulting representation advertises the real feature's exact relevance, given the research focus. Representation as-if and its basis in functional similarity can help account for the insight provided by this false representation of electrons as if they were particles with classical orbits. As Bokulich says, these fictions "correctly capture in their fictional representation real features of the phenomena under investigation" (104).

In my view, the real features captured in this false representation, or fictionalization, are those involved in a particular causal pattern of interest. The sought pattern regards the way in which magnetic and electrical charge interact to yield the absorption pattern displayed by an ion. This is of interest in virtue of being an anomaly from the perspective of quantum theory. The pattern in the magnetic and electrical charges' influence, as represented by closed orbit theory, qualifies as causal on a manipulability approach. Certain interventions to the magnetic charge would change the absorption pattern in a way captured by closed orbit theory, despite—or really because of—its idealizations. Real features of the electrons' probability densities and the interference patterns among them that result from the ion experiencing magnetic charge are responsible for the anomalous absorption pattern. These features are adequately, if falsely, represented by closed orbit theory with a combination of classical and quantum elements—orbits and interference patterns, respectively. A fully quantum representation is unavailable. Even if one were available, it would be exceedingly computationally complex, and crucially, any gains in accuracy would be irrelevant to the focal causal pattern.

Bokulich's (2009) analysis of this case accords nicely with my view of the idealized representation of causal patterns, but there is one difference: Bokulich does not think closed orbit theory captures a causal dependence. She also draws inspiration from Woodward's manipulability approach to causation, but she thinks this reveals relations of counterfactual dependence, which may or may not be causal in nature. In her view, because the orbits are fictions, causal powers cannot be attributed to them. I don't see the difficulty: classical orbits are idealized representations of the changing probability densities of electrons, just as infinite population size is an idealized representation of the actual finite sizes of real populations. In both instances, the idealizations represent real features of the phenomenon of interest as if they were different than they are, in a way that serves the immediate research aim. I think closed orbit theory represents a causal pattern, just like the other cases I have analyzed so far. But I also don't think much importance rides on the difference between our interpretations. As I indicated in Chapter 2, one is free to reinterpret my claims about causal patterns as simply regarding manipulability patterns.

To round out my discussion of this example, notice that this use of idealization occurs in a different kind of scientific context from my earlier examples: in fundamental physics that is supposed to be wholly general. Yet the idealized representation is just as crucial and for similar reasons. Those reasons include (at least) the limits of computational power or human cognitive limits and a limited research focus. Despite the supposed generality of fundamental physics, the nature of the idealization is shaped by the particular research focus. Some other quantum phenomena are better approached with a fully quantum account of electrons, and closed orbit theory is quite limited in its scope. Notice also that, unlike my earlier examples, this idealized representation was developed in response to anomalous experimental data, and its success is judged by its ability to accommodate those data and data about related phenomena.

I conclude this section with a consideration of how my emphasis on idealization to the end of capturing causal patterns relates to research in climate science. This is not a paradigmatic example of physical science, but there are two reasons I want to consider this research. First, climate research provides a nice contrast to all of the research discussed so far. It relies heavily on computational models and apparently does aim for accurate predictions, as I have suggested at least behavioral ecology research into cooperation and studies of human aggression do not. These predictions are not primarily generated to corroborate or disconfirm general theories, as in quantum physics, but are themselves taken to be the main goal. It is well established that anthropogenic climate change has occurred and is continuing. A prominent focus of climate research is now to predict the extent of future warming and the nature and extent of climate effects this will produce. Second, including climate science in a discussion of projects in the physical sciences helps to illustrate the range of types of projects in that arena. There is of course much more to the physical sciences than fundamental microphysics, but fundamental microphysics has historically received a disproportionate amount of philosophical attention. The physical sciences also include investigations of the very large and of the highly specific. Earth's climate is both.

The focus of global climate research is limited to a single, integrated system. This system is expansive, and it is extremely causally complex. A prominent research focus is generating accurate predictions of future states of this system. Winsberg (2010) addresses the value of computer simulations in grappling with such complex phenomena in general, as well as with climate change in particular. As he details, multiple computer models are often developed to tackle a single predictive question. These models may differ in their assumptions, parameter values, and the type of causal processes they take into account. They employ equations that are designed to capture the local weather changes resulting from complex interactions among a broad variety of causes. Some of these causal interactions are relatively straightforward to capture, but others are not well understood or else difficult to represent accurately in a global climate model (Parker 2011). Indeed, Winsberg says that climate models are "a motley mixture of physical theory, approximation, physical intuition, and trial and error" (93). There is thus often little confidence in the accuracy of any one model and little confidence in some of the components in any number of models.

A primary strategy for making and assessing predictions in these circumstances is to ascertain what predictions are robust across a range of models with different assumptions, parameters, or types of representations. This technique is called robustness analysis. The idea is that demonstrating that a result holds across a range of different plausible assumptions and representations of the system indicates that the result is not sensitive to any of those specific assumptions or representational choices and is thus confirmed as a plausible prediction. The use of robustness analysis is a topic of philosophical controversy. Orzack and Sober (1993) criticize the idea that robustness is in general a guide to truth; Weisberg (2006, 2013) responds to this criticism by outlining how, in his view, robustness analysis can provide a form of confirmation without direct empirical support. His analysis, though, is focused on the use of robustness analysis to ascertain what I call causal patterns— how one or a few factors influence a recurring phenomenon of interest. In contrast, Wendy Parker (2011) addresses the use of robustness analysis in climate science in particular, so her treatment is tailored specifically to the aim of accurate prediction. She is critical of robustness analysis's current ability to generate support for climate models' predictive hypotheses.

As Parker (2011) notes, even if she is right about the shortcomings of the use of robustness analysis in climate science, it does not follow that climate policy decisions should be postponed until the science is more certain. This suggests that climate science actually has at least two main aims: first, to generate accurate predictions about future states of the Earth's climate and, second, to guide policy. If today's climate science provides well-confirmed specific predictive hypotheses, those predictions are the best basis for policy decisions. And yet, even if this is so, there may well be further requirements on predictions to enable them to serve the aim of policy guidance. For example, predictions must be made not only about the likeliest outcomes but also the riskiest possible outcomes in order for them to inform responsible policymaking. If Parker's concerns about predictive robustness analysis are on mark, then the two aims of accurate prediction and policy guidance may come apart more significantly. It may be that the research most valuable in guiding policy is of less direct value in generating precise predictions, and the most well-confirmed predictions may not provide the best basis for policy decisions.

The case studies of behavioral ecology and research into human aggression earlier in this chapter have as a central aim the representation of causal patterns. In contrast, climate science's aims of prediction and policy guidance are pursued partly at the expense of accurate causal representation. Robustness analysis enables researchers to set aside causal influences that are difficult to model or estimate accurately. Nonetheless, this research directly capitalizes on causal patterns. The causal pattern of primary interest is the relationship between human influences on climate—most significantly, our carbon emissions—and predicted future scenarios. The computational models are based on scientific knowledge of some of the most important causal influences on the Earth's climate and how human activities affect them. But a full account of the causally complex system, or even of the variety of relevant factors, is much less important than an appreciation for the specific causal roles of the factors over which we humans exert control. Further, for any given model or ensemble of models, the focus is the causal contribution of human impacts to a particular variable, such as corn yield in North America, not to the climate system taken as a whole. What variables serve as the focus are selected in virtue of their interest to scientists, funding agencies, and so on, and they often are of interest because they are relevant to the well-being of humans, our societies, or our economies. So, even though the research aim of climate science is not primarily the representation of causal patterns and still less accurate representation of the entire, causally complex system of the Earth's climate, this research does capitalize on causal patterns, and it focuses on the patterns that obtain between factors we humans influence and variable outcomes that are of particular interest to us.

As in the earlier case studies, idealizations play a pivotal role in climate research. Climate models are complex, computational models that employ equations of different statuses. As Parker (2011) discusses, some of those equations are (idealized) representations of relatively well-understood processes. Others are much rougher—that is, more heavily idealized—representations of processes that are more difficult to represent at the needed scale. Other equations employed may be idealizations in their entirety, in virtue of our limited understanding of the dynamics they are used to stand in for, or else they may be attempts at realistic representation, facilitated by still more idealizations. As Winsberg (2010) outlines, there are decisions to make regarding how to combine these equations into a single model—roughly, in how to represent interactions among the processes the model's equations individually represent. Winsberg makes the case that this "coupling" occurs by a very messy process, which renders the model useful for only a specific predictive task. This process of combining equations into a single model is yet another opportunity for idealizations to facilitate the success of climate models. The primary motivations for many of these idealizations are limited computational power and limited background knowledge, so in theory, de-idealization would be beneficial when possible, to the end of facilitating more accurate predictions. But while individual idealizations may be eliminated, the sheer scope of idealization makes other techniques of managing inaccuracies more useful, including adjusting the idealizations in individual models and robustness analysis across ensembles of models.

I suggested that the confirmation of models of the evolution of cooperation and of human aggression is indirect at best. In contrast, there are a number of forms of empirical confirmation employed in climate modeling, including testing particular parameter values and models' predictions for temperatures and events of past years. Still, this confirmation is not sufficient to identify one or even a few best models, even for a given predictive task. Accordingly, the main predictions must be corroborated indirectly; this helps account for the significance of robustness analysis. The limitations in direct confirmation are particularly interesting in this research, in light of the emphasis on the aim of accurate prediction. In the above discussion of behavioral ecology models of cooperation, I linked the deemphasis of accurate prediction to the failure to empirically confirm the models. I suggested that the limited accuracy of behavioral ecology models is due to the lack of emphasis on applying these models to specific evolutionary phenomena, and the latter is what would warrant specific predictions. In climate research, empirical confirmation can be difficult to come by, but it is highly valued. Any limitations in empirical confirmation are not due to a deemphasis of predictive accuracy or of application to a specific system. Indeed, all emphasis is placed on predictions of the future behavior of a single system: Earth's climate. In this research, limitations in direct empirical confirmation are attributable to one source, namely, the causal complexity of this focal system.

3.4 Surveying the Diversity

Here I briefly spell out the main conclusions about the research projects I have surveyed in this chapter. To begin with, even in the small sampling of scientific research I have described in this chapter, there is an impressive diversity of aims and methods. The aims of these research projects vary from accounting for the causal role of one factor in producing a very general, heterogeneous phenomenon in the case of the evolution of cooperation, to accounting for the action of certain types of factors in the more specific phenomenon of human aggression, to issuing specific predictions and guiding policy decisions in the case of climate science. A large variety of roles are played by data, but direct, empirical confirmation of a prediction or the individual elements of a model are a main emphasis only for closed orbit theory in physics and, toward a different aim, in climate science. Many of these research projects are profitably described as illuminating causal patterns, although this too occurs in different ways. Variations include what types of causal factors are focal; the principles governing the selection of those focal factors; and whether the aim is to illuminate a general causal role, or causal action in more specific, set circumstances, or to treat causal influences in such a way so as to maximize predictive accuracy or policy guidance. In no case was the aim all of these simultaneously. That is, none of the research surveyed aimed to provide a complete accounting of all of the relevant causal influences, to the end of making accurate predictions to inform policy.

It is remarkable that even with all of the differences among the varieties of scientific research I have discussed, all prominently employ idealizations. Yet the nature of these idealizations is relative to the aim of the research, as is most clearly demonstrated by the different idealizations involved in the various approaches to human aggression research. The type and degree of corroboration sought for idealizations also vary. Idealized assumptions in behavioral ecology are sometimes examined to determine their well-foundedness, whereas this is uncommon in the approaches to human aggression research I surveyed. Yet another distinct approach is found in the example of fluid dynamics research, where geometric considerations were used to justify the primary idealizations. The extent of idealization and the unevenness of their corroboration are evocative of my claim in Chapter 2 that

THE DIVERSITY OF SCIENTIFIC PROJECTS

idealization is rampant and unchecked. Yet variation in the nature of idealizations and in the attention they receive accords with my view that, although idealization is rampant and unchecked, it is not unprincipled.

I have of course come nowhere near to cataloging the full diversity of the aims and methods of science. There were three main purposes of this exercise. First, this survey constitutes a step back from the across-the-board generalizations I made in Chapter 2 about the significance of causal patterns. Causal patterns comfortably occupy the role once thought to be played by universal and exceptionless laws of nature. Yet the nature of their significance varies among different scientific projects. Second, this survey shows the extent of idealization in science and the range of roles idealizations occupy. This accords with my view of rampant and unchecked idealization and yet clarifies how the nature of idealizations and the attention they receive vary among different research programs. Third, and last, all of this diversity begins to suggest that science does not proceed in lockstep toward truth, in the hopes of simultaneously generating accurate predictions, corroborating assumptions, and constructing explanations. Something different—and much messier—is going on. This is the topic to which I now turn.

4

Science Isn't after the Truth

In the previous chapter, I considered a few examples of contemporary scientific research. Despite diverse aims, representational tools, relationships to data, and connections to other research, all of those research projects benefit from significant idealization. Behavioral ecology; genetic, neurobiological, and sociological investigations of human aggression; fluid dynamics; quantum physics; and climate science all persist in the use of idealizations. Even in these few examples, the idealizations are present for different reasons and different steps are taken to accommodate them. But in each case, not much de-idealization is attempted. The goals of using idealized models vary, but in none of these examples does the goal seem to be simply attaining a more accurate representation of a specific target system. I argued that the aim of the behavioral ecology research into cooperation is to represent general causal dependencies without representing or predicting the specific features of individual evolutionary outcomes. Research on variation in human aggression is focused on a particular phenomenon, but the different approaches to this research each focus on a small subset of the contributing factors. In the example from fluid dynamics, some features of individual phenomena are also set aside to the end of characterizing a broad class of phenomena. In quantum physics, elements of a rejected theory, classical mechanics, are employed as a fiction to account for anomalous atomic behavior. Finally, for computational models of climate change, the goal is generally accurate prediction, without direct emphasis on accurately representing the full causal structure of the Earth's climate. Because of their continuing idealizations and lack of emphasis on accurate representation of specific phenomena, I believe these examples resist interpretation as aiming for the truth, successive approximation, or increasingly accurate representation of phenomena. Indeed, I suspect this is so for most scientific research.

In this chapter, I develop an account of the aims of science in light of these considerations. In my view, science is not in a lockstep pursuit of truth. Instead, there are a variety of scientific aims that are in tension with one another, and the ultimate epistemic aim of science is not truth but understanding. Truth about phenomena occasionally accompanies the successful pursuit of scientific understanding but not often. Instead, understanding is achieved by revealing causal patterns, patterns that are imperfectly embodied by phenomena and that relate to only some elements of phenomena. My first task in this chapter is to develop the idea that understanding is the epistemic aim of science. I then make the case that, in light of this reinterpretation of science's epistemic aim, science should also be taken to have a number of other distinct aims, aims that are often in tension with one another. I do both of these things in section 4.1. Then, in section 4.2, I explore some implications of my endorsement of understanding as science's epistemic aim. Taking understanding to be a form of epistemic success, and yet suggesting that understanding can be obtained by means other than truth, is on its face an uncomfortable combination of commitments. I address this by further considering the nature of understanding and by discussing how my account is compatible with the idea that science produces knowledge. My view does allow for scientific knowledge, but as will become clear, it is knowledge of a special kind.

At this juncture, it may be helpful for me to remind the reader that my project in this book is not to undermine the success of science but to clarify the nature of that success. This is particularly important to keep in mind in this chapter, for the idea that science does not pursue truth could easily be interpreted as a criticism of science's epistemic value. That is not my intent. To the contrary, I expect that this reinterpretation of science's epistemic aim—as not truth but understanding—will help make clear the great extent of science's epistemic success. Think of it this way. Science is only partially successful at uncovering truths about the world, but it is much more successful at generating human understanding of the world.

4.1   The Aims of Science

In Chapter 2, I argued that there are several intertwined reasons to idealize, that idealizations play a positive representational role, and that there is rampant, unchecked idealization throughout the scientific enterprise. In

Chapter 3, I showed that, at least in the research surveyed there, idealizations are present for different reasons and accomplish different ends, and different steps are taken to accommodate them. Yet they are not eliminated, nor are they minimized. One common view is that all of this idealization may be necessary, and it might be here to stay, but it results in representations that are lacking in various ways. Accordingly, the view goes, we must look for a subsequent step, a way to connect these idealized representations to the successful pursuit of the aims of science, whether those are true theories, successful predictions, complete explanations, or accurate representations. The textbook version of this view would hold that science aims for truth, and so idealized representations must be de-idealized to achieve this aim. It seems that Odenbaugh and Alexandrova (2011), for example, assume something like this view, for they argue that without the removal of all idealizations—complete de-idealization—we have "no ground, beyond that of our background knowledge that informed the model, for claiming that the model specifies a causal relation" (765). Odenbaugh and Alexandrova conclude that even the use of multiple models with different idealizations—called robustness analysis—cannot yield the description of a causal mechanism. Thus, they claim, analysis based on multiple, idealized models does not allow for the confirmation of models, nor can it generate explanations.

Other versions of this view do not hold de-idealization to be necessary but still anticipate the need to bridge the gap between idealized models and the traditional aims of science. Wimsatt (2007), for instance, argues that there are several ways in which idealized, "false" models can be used to produce "truer" theories without recourse to de-idealization. Similar to Odenbaugh and Alexandrova's (2011) concern with causal description and explanation, Rohwer and Rice (2013) argue that at least one purpose of idealizations, the investigation of general patterns across heterogeneous systems, prevents the accurate description of causal factors and thus the formulation of explanations (although they emphasize that the resultant model can still be explanatory). These authors all endorse the continuing practice of idealization, but they also hold idealized models to be somewhat removed from achieving the traditional aims of science. They explicitly or implicitly commit themselves to an intermediary step of some kind between idealized representation and the achievement of science's aims. On this strategy, even though idealized models are of scientific value, they are not sufficient to provide adequate explanations, trustworthy predictions, causal representations, and so on—or at least not by themselves.

One could instead take the opposite approach to reconciling idealization with the aims of science. The observation of widespread idealization in science, and the distance between idealized models and traditional articulations of the aims of science, might be seen as grounds for concluding that those traditional articulations of the aims of science are incorrect. On this alternative approach, nothing has gone wrong with or is lacking from idealized models, and no intermediary step is needed for idealized models to achieve the aims of science. Those aims just stand in need of clarification. This is the tack I take here. I motivate the idea that, in an important sense, science is not after the truth. Much of science resists interpretation as successive approximation or increasingly accurate representation of phenomena, to the end of explaining and predicting those phenomena. As a result of rampant, unchecked idealization, many of the best products of science are not things we believe to be true.

4.1.1 UNDERSTANDING AS SCIENCE'S EPISTEMIC AIM

Wimsatt (2007) points out, regarding idealized models, that "unless they could help us do something in the task of investigating natural phenomena, there would be no reason for choosing model building over astrology or mystic revelation as a source of knowledge of the natural world" (101). This must be right. Idealized representations, even though they are false in some regards, must get us somewhere that astrology or mystic revelation does not. At issue is what exactly idealized representations help us accomplish and, in particular, the nature of their epistemic value. This question is made even more significant if I am right that idealizations are rampant and unchecked in science. Here I will explore the idea that false models—and idealized representations more generally—are not a means to truer theories, as Wimsatt believes, but are instead themselves directly responsible for achieving the aims of science, including its epistemic successes. Because idealizations are patently untrue, their continued presence in models and other representations cannot be justified by their contribution to the truth of those representations. Accordingly, if idealizations directly contribute to science's epistemic aim, that epistemic aim must be something other than truth.

A first step toward a conception of the epistemic aim of science to which idealized models can directly contribute is provided by Elgin (2004). Elgin is also impressed by how many scientific laws, models, and theories diverge from the truth in various ways. Her aim is thus to show how these scientific products can be epistemically valuable without being entirely true. She says,

> I take it that science provides an understanding of the natural order. By this I do not mean merely that an ideal science *would* provide such an understanding or that in the end of inquiry science *will* provide one, but that much actual science has done so and continues to do so. (114)

Elgin's strategy is to accept today's actual science as a successful venture, then look to see what this science accomplishes. This is reminiscent of the approach I take in this book, as outlined in Chapter 1. What Elgin finds is that science regularly produces understanding, even as it falls short of producing truths. Rather than make excuses for the myriad ways in which our scientific products fail to be true, Elgin suggests that science's epistemic success consists of understanding, not (necessarily) truth. I believe this is the right way to interpret idealizations' epistemic contribution. Idealizations cannot directly contribute to science's epistemic success in virtue of their truth, but they might in virtue of their contributions to human understanding.

For this approach to have promise, two requirements that seem to be in tension with one another must be met. It must be possible for understanding to be furthered by things that are not true, but understanding must still qualify as an epistemic success. The concept of understanding has a key feature that makes it possible to navigate this tension: it has a dual nature. Understanding is at once a cognitive state and an epistemic achievement. Because understanding is a cognitive state, it depends in part on the psychological characteristics of those who seek to understand. This distinguishes understanding from truth, for whether a proposition is true in no way depends on the psychology of one who entertains or believes that proposition. Understanding's nature as a cognitive state provides an avenue for idealizations to make a positive contribution to it, despite their lack of truth. On the other hand, because understanding is also an epistemic achievement, it is subject to standards for success. Understanding is not any old "aha" moment; it is not merely a psychological state. Put another way, a felt sense of understanding is not sufficient for the possession of genuine understanding. Genuine understanding also requires successful mastery, in some sense, of the target of understanding. Both Grimm (2006, 2011, 2012) and Strevens (2013) describe this mastery as a form of grasping; as Grimm (2012) stresses, "grasping" is a success term. For Grimm, one cannot have objective understanding without possessing truth, so he must take truth to be one of the standards governing successful understanding. But one might reject truth as a requirement for understanding while still maintaining that there are standards governing whether genuine understanding has been obtained.

Let me address in greater detail both how idealizations might contribute to understanding qua cognitive state and then also how those contributions might be accommodated by the standards governing understanding qua epistemic achievement. For both tasks, I will again take inspiration from Elgin (2004). Consider first the question of idealizations' contribution to understanding. Elgin suggests that "felicitous falsehoods," or idealizations, can facilitate understanding insofar as they "impose an order on things, highlight certain aspects of the phenomena, reveal connections, patterns and discrepancies, and make possible insights that we could not otherwise obtain" (127). She gives the simple example of drawing a smooth curve and treating the data's deviation from the curve as error or noise. This notion of how idealizations can contribute to understanding should bring to mind my emphasis on the centrality of causal patterns in science. Recall that causal patterns are regularities, perhaps with deviations and exceptions, embodied by some limited range of phenomena. These patterns qualify as causal on Woodward's manipulability approach to causation. In my view, much of the value Elgin attributes to idealizations relates to causal patterns and how they are embodied by phenomena. Depicting a causal pattern highlights certain aspects of a phenomenon. It imposes an order when the pattern is only imperfectly embodied and thus enables insights about the pattern we would not have access to simply by considering the phenomenon. Further, "connections, patterns, and discrepancies" are all elements of the relationships among phenomena embodying a pattern—and the outliers that constitute exceptions.

I argued in Chapter 2 that depicting causal patterns regularly motivates departures from accuracy of any given phenomenon; this is why idealizations are used to represent as-if. Put in these terms, the present idea is that idealizations positively contribute to generating understanding by revealing causal patterns and thereby enabling insights about these patterns that would otherwise be inaccessible to us. These are contributions to understanding as a cognitive state, that is, to the felt sense of understanding. The idea that revealing patterns has a special role to play in generating human understanding is corroborated by research in psychology. Williams et al. (2013), for instance, provide empirical support for the idea that human understanding is furthered by generalizations about broad patterns.[1]

This brings us to the question of how this contribution to understanding might be accommodated by the standards governing understanding qua epistemic achievement. Idealizations are departures from the truth, so if truth is required for understanding, it seems that idealized representations may be barred from generating genuine understanding or, at the very least, that idealizations themselves cannot promote genuine understanding. However, as I suggested above, one might take there to be requirements for the epistemic achievement of genuine understanding but deny that one of those requirements is truth. Elgin (2004) does just this. She proposes that truth is a requirement on understanding but that it functions as a threshold requirement. For Elgin, a claim must be "true enough" to be epistemically acceptable; that is, any divergence from the truth must be negligible, or safely neglected.

She argues that whether a claim is epistemically acceptable depends on the function it plays in an argument, explanation, or theory—or, one might further generalize, in a model or any other scientific representation. Elgin appeals to the example of Snell's law, which governs light's angle of refraction when it passes from one medium to another. Snell's law is only true of optically isotropic media, but it is true *enough* of media that are nearly isotropic, which include a wide range of media in which physicists are interested.

I think something like Elgin's threshold requirement of "true enough" must be one of the standards for genuine understanding. But in my view, whether a claim is true enough cannot depend only on its function in an argument, explanation, or theory, as Elgin holds. It also depends on the purpose of the research to which it contributes. Consider that, as Elgin points out, Snell's law applied to anisotropic media is of limited use "if we are interested only in the path of a particular light ray" but is useful "if we are interested in optical refraction in general" (118). This divergence in whether Snell's law generates understanding cannot be accounted for by the role played by the assumption of isotropic media, for that is the same in the two instances. What varies are the researchers' interests and thus what they intend to use the law for. So, in my view, whether a claim can help generate genuine understanding must depend also on the precise aim of the research. This is only a minor revision of Elgin's idea, but it hints at a more significant shortcoming of her conception of understanding, which I address below.

When the epistemic achievement of understanding is distanced from truth in this way, it also must be distanced from belief. Otherwise, one would be in the position of endorsing the epistemic value of believing false claims, which is counterintuitive at best. And indeed, identifying an assumption as an idealization is tantamount to declaring it to be false, that is, not to be believed. No biologist actually *believes* she is dealing with an infinite population of organisms or (in many cases) organisms that reproduce asexually. This leads to the recognition that, as Elgin (2004) puts it, "understanding is often couched in and conveyed by symbols that are not, and *do not purport to be*, true" (116, emphasis added). Likewise, Winsberg (2010) argues that the proper attitude to have toward "successful model-building principles," which I would call idealizations, is not holding them to be true or even approximately true (134).[2]

Distancing false claims that help achieve understanding, a form of epistemic success, from belief is made simple when it is appreciated that a variety of cognitive attitudes can be significant to science. Elliott and Willmes (2013) argue that scientists adopt a number of different cognitive attitudes other than belief toward the products of science. They follow Cohen (1992) in distinguishing between accepting and believing a body of content, such as a hypothesis, theory, model, or other representation. They define acceptance as follows: S accepts that h iff S presupposes h for specific reasons in her deliberation. Given this formulation, different forms of acceptance correspond to different reasons for presupposing. Two of Elliott and Willmes's examples are accepting as worthy of further investigation and accepting as the basis for policymaking. Neither of these forms of acceptance is a sufficient commitment to be an element of understanding. Understanding, as an epistemic state, requires *epistemic acceptance*, that is, presupposing for epistemic reasons. Conceived in this way, epistemic acceptance includes a wide variety of levels of commitment, ranging from assuming contrary to fact, to taking to be accurate enough for present purposes, to believing as true. What unites this range of attitudes is that, in the right circumstances, each is a sufficient commitment toward a posit for it to contribute to the achievement of understanding. On this view, contributors to understanding, including idealizations, must be epistemically accepted—just as they must be epistemically acceptable, or what Elgin calls "true enough"—but they need not be believed. Indeed, idealizations can be epistemically accepted simply when they are assumed contrary to fact.

This conception of understanding as an epistemic success that does not require full truth is a promising first step toward accounting for how idealized representations and idealizations themselves can be of direct epistemic value. Idealizations fit the bill as Elgin's "felicitous falsehoods [that] figure in cognitive discourse not as mistaken or inaccurate statements of fact, but as fictions" (123). In the proper representational roles and as part of the right research projects, idealizations are epistemically accept*able*, and they are epistemically accept*ed* as posits assumed contrary to fact. I have suggested that idealizations contribute to understanding by representing as-if to the end of depicting a causal pattern, thereby highlighting certain aspects of that phenomenon (to the exclusion of others) and revealing connections with other, possibly disparate phenomena that embody the same pattern or, in some cases, that are exceptions to that pattern.

This view sets up understanding as a proper alternative to truth as the epistemic aim of science, and it demonstrates how idealizations can directly contribute to the aim of understanding. Idealizations facilitate understanding, but not truth, when they enable the representation of cognitively valuable connections and patterns that more accurate portrayals would miss. This is a departure from a truth-centric conception of science and a move toward a conception of science's aims that better accords with rampant and unchecked idealization. However, the view I have developed so far requires two amendments. The first is that so far I have focused exclusively on a purely epistemic aim of science—namely, understanding—but science has a number of distinct aims. Some are epistemic, such as understanding; others are nonepistemic, such as action within a short time span; and still others seem to involve both epistemic and nonepistemic elements, such as accurate prediction (see Elliott 2013). Idealization makes a purely epistemic contribution to the production of understanding, but it likely contributes in different ways to different aims. Instead of a single successor aim for truth to which idealizations contribute, we should thus expect a variety of scientific aims, aims that may be facilitated by idealizations in different ways. The list of science's aims may even be open-ended, for science is a continually creative process; its procedures as well as its products are always in development. Nonetheless, I think some generalizations can be made about these aims. First, many or all relate in one way or another to establishing and employing causal patterns, as I set out in Chapter 2 and illustrated in Chapter 3. And second, all of these aims further human cognition, action, or both, as a science by and for human beings should. I say more about these aims and idealization's roles in promoting them in the next section.

The second amendment to the view I have sketched so far is a revised requirement of epistemic acceptability. So far I have based this requirement on Elgin's (2004) threshold requirement of "true enough," but this preserves a role for truth that is both too central and too uniform. As a result, it does not accommodate all the ways in which idealizations facilitate understanding. To see the problem, consider what Elgin says about the epistemic value of the ideal gas law:

> We understand the properties of real gases in terms of their deviation from the ideal. In such cases, understanding involves a pattern of schema and correction. We represent the phenomena with a schematic model, and introduce corrections as needed to closer accord with the facts. (127)

Phrasing this as schema and correction to "closer accord with the facts" sounds awfully similar to the traditional view of idealization's contribution outlined above, according to which idealizations are distortions, to be overcome or circumvented in the pursuit of truth. But the use of idealized representations like the ideal gas law and Snell's law (mentioned above) often does not conform to this picture of schema and correction. In many circumstances, the idealized representation alone, without any corrections, is employed. For example, the ideal gas law remains central, even though there is a more accurate alternative—namely, the van der Waals equation, which takes into account molecular size and intermolecular attraction, properties the ideal gas law simply idealizes away.

Part of the difficulty here is Elgin's presupposition that all of science generates claims that figure into arguments, explanations, or theories. This is not so. Scientific modeling may proceed largely independently from theory, and there is a diverse array of scientific products, many of which have little or no relationship to theory or explanation. A broadened conception of scientific products undermines the idea that these products always benefit by more closely according with the facts. Elgin (2014) also claims of the ideal gas law that "if, for example, evidence shows that friction plays a major role in collisions between gas molecules, then unless compensating adjustments are made elsewhere, theories that model collisions as perfectly elastic spheres will be discredited" (129). This is presumably because the claim that gas molecules are perfectly elastic spheres would be revealed as not true enough. Maybe this would be so for *theories* of gases that employ that assumption. But Elgin doesn't seem to appreciate that this would not discredit all representations that rely on the posit that gas molecules are perfectly elastic spheres. There would undoubtedly still be perfectly good reasons, including epistemic reasons, for continuing to model gas-molecule collisions in this way. This would be just like biologists modeling a population as if it were infinite, even though it is relatively small, when their focus is the role of natural selection. Or like physicists appealing to an element of classical mechanics, a discredited theory, to account for quantum behavior. Both of these projects would be hindered, not furthered, by closer accordance with the facts.

Elgin's proposed standard for epistemic acceptability is thus still too truth-conservative. Whether a posit is epistemically acceptable does depend on its role in a representation, but representations are developed for many different purposes. Epistemic acceptability thus also depends on another factor: scientists themselves and, especially, their particular research interests. Representing a population as infinite in size, a gas as comprising noninteracting point particles, or an electron as following a classical orbit can be epistemically acceptable even if radically incorrect when researchers want to understand certain other features of these phenomena. These idealizations are epistemically unacceptable, though, when researchers want to understand, respectively, the role of genetic drift, intermolecular attraction, or electrons' movements. Other considerations about scientists themselves may also be relevant to epistemic acceptability, including their cognitive faculties, psychological characteristics, temporal and spatial location, and any other features that can affect what scientists attempt to accomplish with a representation.

I thus propose the following revised requirement of epistemic acceptability:

> A posit is **epistemically acceptable** when its divergence from truth is insignificant, taking into account (a) the posit's role in the representation and (b) the epistemic purpose to which that representation is put.

Epistemic acceptability of this kind is required for a posit to promote genuine understanding, an epistemic achievement. It may be helpful for me to work through a couple of the examples at hand, labeling the elements (a) and (b) from the above statement of the requirement. The posit that electrons follow classical orbits is epistemically acceptable as (a) an idealization in (b) the closed orbit account of electron absorption when an atom experiences magnet force. This posit approximates some features of quantum electrons in a way that accounts for the behavior in question. But the same posit is not epistemically acceptable as (a) a claim of (b) classical mechanics, taken as a theory, for there is clear evidence that electrons are not discrete entities following such orbits. The posit that a population of organisms is infinite in size is epistemically acceptable as (a) an idealization in (b) an account of the role of natural selection in bringing about some evolutionary outcome, even if the population is rather small. But, if the population is rather small, it is not epistemically acceptable as (a) an idealization in (b) an account of the predominant evolutionary influences. The degree of divergence from truth consistent with epistemic acceptability is, in turn, a guide to what form of epistemic acceptance to expect; the main form of epistemic acceptance for idealizations is posited contrary to fact.

Because understanding is also a cognitive state, it's fitting that its achievement depends in part on the features of epistemic agents. The relevant features include, at least, what they seek to understand (determined by their research interests), what they are in the position to understand (determined in part by their cognitive faculties, background information, and temporal and spatial location), and what contributes to their understanding (determined in part by their cognitive faculties, background information, and psychological characteristics).[3] All of these features contribute to what exactly scientists attempt to accomplish with representations and, accordingly, the level of divergence from the truth that is consistent with epistemic acceptability. In other words, all of these features contribute to shaping what promotes genuine understanding. This formulation of the requirement of epistemic acceptability thus renders understanding qua epistemic achievement consistent with understanding qua cognitive state.

With these two amendments, this account of understanding as science's epistemic aim successfully accommodates the direct epistemic contributions of idealizations. First, because science has a variety of aims, understanding may replace truth as the primary epistemic aim, but it is not the primary aim of science simpliciter. Accordingly, the aim of understanding does not alone account for all uses of idealizations, just their narrowly epistemic value. And, second, with the epistemic aim of understanding, the research purpose to which an idealized representation is put must be taken into account to determine whether an idealization promotes understanding. Idealizations, no matter how little they resemble the systems they represent, may in the proper circumstances facilitate an understanding of those systems.

A significant implication of this view is that an idealization can be radically untrue but nonetheless facilitate understanding. The divergence from truth can contribute to representing a pattern embodied by the focal phenomenon but from which the phenomenon deviates to some degree—perhaps even a large degree. This is the case when, for instance, an infinite population is assumed when the population size is rather small and the causal contribution of genetic drift significant, resulting in an evolved outcome that deviates from what is expected. Such a scenario is a regular occurrence for the behavioral ecology models of cooperation discussed in Chapter 3, which focus on patterns in natural selection's causal role in evolved cooperative behavior. This epistemic contribution of idealizations exists because, as I claimed above, humans judge broad patterns to have great explanatory power. This might be so for any reason; it could simply be an accidental feature of our cognitive processes. Nonetheless, because broad patterns facilitate human understanding, if the primary epistemic aim of science is human understanding, idealizations' role in representing causal patterns is an epistemic contribution.

Other philosophers have similarly stressed idealizations' contribution to highlighting patterns and connections among disparate phenomena, including several whose views about idealization I have already discussed (e.g., Batterman 2002a, 2009; Bokulich 2009; Strevens 2008b). But those accounts develop this idea in a mind-independent way, focusing on how disparate phenomena can exhibit the same high-level patterns. For example, in his treatment of idealization in fluid dynamics, Batterman (2009) emphasizes that fluids at the point of breakup into droplets constitute a universality class, characterized by the idealized representation. In contrast, on the view of idealizations' contribution to understanding developed here, that contribution depends both on general human cognitive traits and researchers' particular cognitive traits, including their research interests. Idealizations' epistemic contribution consists in their usefulness in depicting causal patterns. But broad causal patterns promote human understanding because of facts about our cognition, and which causal patterns promote understanding depends on researchers' interests, background information, location, and so on.

Let's take stock. So far I have motivated the idea that idealizations can be of direct epistemic value by positively contributing to human understanding. This is different from truth. Idealizations cannot be true or approximately true, but they can be epistemically acceptable, and they can—and regularly do—promote human understanding. So, *if* understanding is a central epistemic aim of science, then idealizations can make direct epistemic contributions to science. My last task in developing a view of the epistemic aim of science to which idealizations can directly contribute is to motivate the antecedent of that conditional. Even if I am right that idealizations can contribute to understanding and that understanding is an epistemic achievement, why think understanding is a central epistemic aim of science? Let me rephrase this as a contrastive question about understanding versus truth, setting aside for now the question of whether there are additional epistemic aims besides these two. Why think that understanding, and not truth, is the epistemic aim of science?

Two ideas I developed in earlier chapters are relevant to answering this question. First, I have argued that our scientific enterprise has been indelibly shaped by its human practitioners and audience. Our scientific products are designed to facilitate human cognition and action. Taking the epistemic aim of science to be understanding helps to account for this human influence in a way that is not epistemically problematic. Scientists' cognitive characteristics and interests can never influence what is true, but these can shape what generates understanding. Second, I have argued that idealization is rampant and unchecked in science. Indeed, heavily idealized representations are regularly chosen even when less idealized representations are available. This would be somewhat mysterious if the epistemic aim of science were truth, for idealizations always detract from truth. But if the epistemic aim is instead understanding, this provides an epistemic reason for sometimes preferring more idealization to less. I thus posit understanding, and not truth, as the epistemic aim of science.

This view might find additional support from a strand in epistemology that defends understanding as the goal of inquiry in general. Jonathan Kvanvig (2003) proposes that understanding has epistemic value that knowledge does not. He takes understanding to differ from knowledge, the traditional epistemic goal, by coming in degrees and in not requiring justification. For Kvanvig, understanding is still factive, that is, it does require truth. Elgin

(2004, 2007) disagrees; as we have already seen, she thinks understanding does not require full truth. Linda Zagzebski (2001) similarly focuses on the epistemic value of understanding, but she claims that what distinguishes understanding is nonpropositional representation. Finally, Grimm (2006) argues that understanding is not distinct from knowledge but a species of it. These authors all emphasize the significance of understanding as an epistemic achievement. For most, an element of that significance stems from what I have called the dual nature of understanding, as both cognitive and epistemic.

For my purposes, the question of whether understanding is factive, thus requiring truth, is of central importance. I have suggested that deviations from truth can positively contribute to understanding—that is, less truth can, in the proper circumstances, lead to greater understanding than would more truth. Notice that this might be distinctive of science; it may be that understanding is not by and large furthered by falsehoods. My contention is that within science it frequently is. Thus, in science, truth plays only a supporting role. When scientists do aim for truth, this is only because the circumstances make it so that truth is the best way to promote understanding. This is unusual in science in virtue of the complex world under investigation and the value of causal patterns to human understanding.

Taking understanding to be the epistemic aim of science would account for the value of many of the intertwined reasons to idealize identified in Chapter 2 and summarized in Table 2.1. What I have said so far about idealizations' contribution to understanding by representing causal patterns motivates permanent reasons to idealize in virtue of causal facts and scientists' specific research interests. Idealizations can also contribute permanently to understanding in virtue of computational limits—for example, by enabling equations with analytical solutions (Morrison and Morgan 1999; Batterman 2002a; Elgin 2004), cognitive limits, and pedagogical usefulness. The pedagogical value of some heavily idealized representations accounts for the crucial role of idealized models in science textbooks, as Giere (1988) and others have emphasized. All of these reasons to idealize can be thought of as influencing which causal patterns are focal. To return to the example of computational limits, a causal pattern from which the phenomenon deviates somewhat significantly may nonetheless be of particular interest if that focus enables an analytical solution.

These and other permanent reasons to idealize are of enduring value to human understanding, but temporary reasons to idealize can facilitate understanding at some particular stage of scientific inquiry. An idealized model may facilitate understanding because of a limitation in our current computational powers, in virtue of what modeling approaches happen to be familiar to the researchers, or because some technique happens to get initial traction with a phenomenon. Such temporary reasons to idealize do not justify permanent departures from the truth, so they are easier to assimilate to the traditional view that science aims for truth or increasingly accurate representation. But the idea that science aims to provide understanding provides epistemic grounds for all of these many reasons to idealize and suggests something all those reasons have in common. All can be steps toward representing a causal pattern that is enlightening.

Before moving on, let me forestall an initial concern one might have with taking science's epistemic aim to be understanding, especially understanding that involves departure from truth. One may worry that this is an overly subjective standard for epistemic success in science. As I have indicated, there is a subjective element to my account. I have argued that what best facilitates understanding depends in part on features of the practitioners and consumers of science, in general and in particular instances, including especially their cognitive requirements and interests. But this element of subjectivity is unproblematic. It is to be expected of a scientific enterprise that has been developed by limited and historically located human beings. What best meets human requirements—including our epistemic requirements—depends in part on the features of humans. Other, more problematic forms of subjectivity do not apply. First, one need not rely on the subjective experience of an "aha!" moment, that is, the felt sense of understanding (Trout 2002), to judge whether a representation facilitates understanding. Instead, objective reasons guide that judgment. The representation must contribute to the relevant humans' sense of understanding to count as understanding qua cognitive state, but all posits in a representation also must be epistemically acceptable for this to qualify as an epistemic achievement. Further, the judgment of whether an idealization increases understanding is based on objective features of the world, including of scientists themselves. This is recognizable for each of the reasons to idealize I have discussed. Limited computational power and absolute computational limits, the limitations of human powers of cognition and the limitations in one researcher's training, and the existence of a causal pattern and the existence of specific research interests: all are objective features of the world under investigation or the scientists leading that investigation that together determine what best generates human understanding.

4.1.2 SEPARATE PURSUIT OF SCIENCE'S AIMS

I have argued that the epistemic aim of science is not truth but understanding. Yet, as I have already acknowledged, there are actually a variety of aims of science, both epistemic and non-epistemic. Traditionally appreciated aims include (at least) prediction, explanation, and accurate representation. Other aims of science have recently received increasing attention, such as providing information to guide policymaking (Douglas 2009b; Parker 2011), enabling action within a short time span (Elliott 2013), and facilitating the public uptake of scientific knowledge (Elliott 2011). There are surely many other aims and many other articulations of these and other aims. My project here is not to delineate the range of scientific aims but to examine the relationship among them. I will motivate the idea that the successful pursuit of one aim generally occurs at the expense of other aims. Accurate prediction is achieved by tools poorly suited to explain, the aim of quick action is at odds with full causal representation, and so on. At root, this is because the different aims of science are furthered by different means, including different idealizations.

Notice first that the diversity of scientific aims is linked also to a diversity of cognitive attitudes toward the products of science. Recall from above that embracing understanding instead of truth as the epistemic aim of science requires shifting from a focus solely on belief to the broader concept of acceptance. This is because many contributors to understanding are not things we believe to be true. I followed Elgin in taking the alternative, broader concept to be epistemic acceptance. But acceptance, like the aims of science themselves, comes in many varieties. Elliott and Willmes (2013) define the acceptance of some posit h as presupposing h in deliberation for specific reasons. From this, I derived the subspecies of epistemic acceptance by restricting the allowable reasons to epistemic reasons. Other varieties of acceptance similarly derive from other, specific aims of science. For instance, acceptance as predictively useful amounts to presupposing for purposes of predictive value (alone). It is clear that one form of acceptance need not entail another. The value of an assumption in the production of understanding, for example, suggests nothing about its predictive usefulness or vice versa.

Just as one form of acceptance need not entail another, the pursuit of one aim of science need not contribute to other aims. Believing a true claim might also be useful for a variety of other aims, such as prediction or guidance of action. In contrast, accepting a claim for its contribution to a specific purpose, even an epistemic purpose like understanding, may not yield similar success with other purposes. Indeed, something stronger is true: success with one aim often *inhibits* success with other aims. Science as a whole employs a variety of tools to achieve, e.g., predictions, explanations, causal representations, a basis for action, and so on. What suits a tool to further one of these aims does not in general well suit it for the other aims. For example, one method used to generate predictions is the analysis of a variety of models with competing assumptions, called robustness analysis. In Chapter 3, we saw that this is a common method in climate modeling for generating accurate predictions. But none of those models is expected to accurately represent the full range of causal influences on climate or to generate understanding of particular features of interest. The core influences that are incorporated into these models are already well understood, and accurate representation of other causal influences is impossible by this method, for the group of models together represents those influences in multiple, incompatible ways. So, in this example, the tool of robustness analysis and these multiple models are in service of one aim and do not contribute to others. A second example of conflicting aims in climate modeling is that the aim of accurate prediction calls for assumptions that are expected to deliver the most likely outcomes, whereas the aim of policy guidance might instead motivate placing particular value on the riskiest possible outcomes.

This division in the pursuit of different aims of science is due in part to the widespread use of idealizations and the variety of purposes to which idealizations are put. Recall from above that whether an idealization furthers understanding depends on the specific goals of the research. This is why, for instance, Snell's law is appropriate to apply to anisotropic media when researchers want to understand optical refraction in general but not when they want to understand the path of a specific light ray. There is a more general corollary to this. Whether an idealization furthers *any* given aim of science depends on the specifics of that aim. Snell's law may help us understand optical refraction, but it is too idealized to give precise predictions of light's path of travel in anisotropic media. For that we need a different tool. And, in the example just above of the use of robustness analysis in climate science, the reverse is the case. Multiple climate models with conflicting assumptions may together be predictively useful, but they do not directly contribute to our understanding of the mechanics of climate change. Idealizations are not believed to be true; instead, they are accepted (in different ways) for different purposes. In light of the centrality of idealizations to science, this results in the splintering of science's diverse aims.

The idea that different aims of science must be pursued separately is evocative of Cartwright's (1983) study of how laws are either false and explanatory or else predictive (but not both). Elsewhere I have defended the view that the aims of explanation and model confirmation push in different directions (Potochnik 2010a). Bokulich (2013) also argues that predictive power and explanatory insight are best provided by different scientific models, even of the same phenomenon, and that less predictive accuracy should thus be expected of explanatory models. Finally, Alistair Isaac (2013) argues that models have specific functions, such as prediction and use in policymaking, and that many of these functions are best satisfied by abandoning the goal of realistic representation. The separate pursuit of different scientific aims is also related to the view, introduced by Richard Levins (1966), that there are trade-offs among the desirable features of models, such as their generality, precision, and realism or accuracy. This is the idea that increasing a model's generality, for example, is achieved by decreasing its precision or accuracy. Different features of models, such as greater generality or greater precision, will differently position those models to contribute to particular representational or predictive aims. In my view, the nature of the selected trade-off reflects the purpose to which a model or other scientific product is put, that is, the aim of science it serves.

Indeed, the tension among aims is even more extreme than I have indicated so far. Not only do the aims of science conflict, but so too do deployments of a single aim. We have already encountered this with the epistemic aim of understanding. Recall from above that idealizations can facilitate understanding of a phenomenon by standing in for features, including causally important features, that are unimportant to the focal causal pattern. Which causal pattern is focal, and thus which causal pattern provides understanding, depends on a researcher's specific interests. So, for example, an evolutionary game theory model may demonstrate the role of natural selection in producing cooperative behavior while occluding the role of nonselective and nonevolutionary influences. These might include specific genetic influences and alternative, nonevolutionary influences like learning, to name just two possibilities. In the context of other research programs, such as those with a population genetic focus or an epigenetic focus, understanding other influences on cooperative behavior will move to center stage, and the game theory model will no longer generate understanding. This shows how the epistemic aim of understanding can, in itself, motivate different scientific products that are appropriate for different research focuses. The same is true for other scientific aims. Different models or ensembles of models of a single phenomenon often are suited to different predictive tasks. Different explanations of the same phenomenon are called for in different research contexts (Lewis 1986).[4]

Table 4.1 depicts how, for the examples of evolved cooperation and climate change, the specific aim and specific deployment of that aim influence what scientific product is called for. For the example of investigating an evolved cooperative outcome, only two deployments of a single aim (understanding) are shown. For the example of investigating climate change, two different aims are shown, with one deployment of each. The ellipses indicate that other aims, deployments of aims, and products exist—indeed, if I am right, many other.

TABLE 4.1. An illustration of how scientific products vary according to what immediate aims and specific deployments of those aims they serve, focused on the examples of evolved cooperative behavior and climate change.

Phenomenon	Evolved cooperative outcome	Global climate change
Aim	Understanding	1. Prediction
	...	2. Understanding
		...
Deployment	1. Role of natural selection	1a. Average temperature in 2100
	2. Genetic dynamics	...
	...	2a. Causal role of land use
		...
Product	1. Evolutionary game theory	1a. Robust finding
	2. Population genetic model	...
	...	2a. Terrestrial biosphere model
		...

These aims, deployments of aims, and products that best serve them are all familiar from discussion just above, except (2a) under global climate change. When the contribution of human land use is focal, an understanding of climate change might be provided by terrestrial biosphere models, which focus on the relatively minor contributor of how changes to land influence climate change.

There are two main reasons for the tension among different aims of science and different deployments of a single aim—reasons, also, for the variety of idealizations' contributions to the aims of science. These are the complexity of phenomena of scientific interest coupled with the limited powers of human cognition and action. Science's diverse aims are all valuable for their ability to further human comprehension and control of the complex world we inhabit. Their furtherance is thus relative to the limitations of human cognition and action. We have seen many examples of how the limited powers of human beings and especially of our cognition, when faced with incredibly complex phenomena, are accommodated with the help of idealizations. But these idealizations are specific to their purposes. This requires focus on one particular scientific aim (at a time), and one particular deployment of that aim, to the exclusion of others. The result is a divide-and-conquer approach, where scientific products are tailored to their precise aims. To successfully predict, represent a certain element of the causal structure, or provide quick guidance for policy, one must sacrifice other goals.

It is instructive to consider the nature of the position most directly opposed to my view of the conflicting aims of science. The opposed position is that the same scientific products individually further all aims of science—at least accurate representation, explanation, and prediction, and possibly also including more human-centric aims like providing grounds for policymaking. If this were the case, a single, best representation would simultaneously offer the best causal representation, the best explanation, and the best predictions. I suspect this kind of view is motivated by a commitment, perhaps implicit, to the idea that scientific products are designed to be true depictions of the world or at least are benefited by being as close to true as possible. If that were generally the case, then one could expect individual scientific products to play all these roles. Representations that were true in all important regards would provide a sufficient causal representation, be explanatorily unimpeachable, and ground accurate predictions. But science does not aim for such truth, or so I have argued. The alternative epistemic aim of understanding gives rise to scientific products tailored to their specific epistemic duty and not well suited to other roles. The aims of science are accordingly in tension.

This view of opposition among scientific aims also conflicts, although less directly, with views that relate confirmation and prediction to explanation. One example of such a view is the idea of inference to the best explanation, where explanatory considerations are used to guide what one believes. Some versions of this idea might accord quite well with the views I develop in this book. What I resist, though, is the idea that an explanation's success is grounds for accepting its *truth*. I have suggested that not truth but understanding, a different form of epistemic success, is in play and that understanding can be directly promoted by untruths. An apparent explanation may be reason to infer that its posits are epistemically acceptable, but in my view, what that amounts to varies. For a vivid illustration of how truth should not be inferred from explanatoriness, recall that the explanation of electron absorption under magnetic charge appeals to classical electron orbits (Bokulich 2009). I have suggested that, given the specific epistemic goal at hand, this posit is epistemically acceptable. But it is certainly not true. Another conflicting view is Heather Douglas's (2009a) idea that the aim of prediction yields insight into the value of scientific explanations. In contrast to this kind of view, I have suggested that the aims of prediction and explanation are actually opposed and that we should expect different scientific products to serve them.

This conception of the relationship among scientific aims helps make sense of a feature of science that would otherwise be puzzling. Across science, there is a proliferation of different approaches to studying the same phenomenon, many of which apparently conflict with one another. We saw an example of this in Chapter 3's discussion of research into human aggression. The different research approaches I surveyed do not combine to create a seamless, integrated account of variation in human aggressivity. As a consequence, those approaches may seem to conflict. A few of the approaches, most notably molecular genetics and social-environmental research, apparently fall on one side or the other of the "nature versus nurture" divide. And Longino (2013) explores in great depth the deep divides among all of these approaches. The tension among different scientific aims and among specific deployments of aims can account both for this proliferation of approaches and for the apparent conflict among approaches. In the case of research into human aggression, if accurate representation of the full suite of causes of aggressivity were the uniform aim, then different research results should be reconcilable or else one or more must be wrong. We would expect that the research would progress toward a single, more complete account of human aggression. In contrast, on the present view, each of these approaches persists because each accomplishes something different. Despite their shared focus on human aggression, each approach has a distinct aim. On my interpretation, many of the approaches aim to provide understanding of some element of human aggression by depicting a causal pattern it embodies. Other aims for this particular research might include predictive goals or use in policymaking. This variety of different aims and deployments of aims generates different research results. Those results may seem to conflict, but instead they simply aim to accomplish different goals.

The proliferation of different approaches to investigating the same phenomenon is sometimes accompanied by persistent, deep disagreements about fundamental principles, including within otherwise functional fields of research. Witness the longstanding, complicated, and divisive debate concerning the significance of "nature" versus "nurture" for human behavioral traits and how they interact (Tabery 2014). One deep disagreement within the field of population biology regards the significance of natural selection versus other evolutionary influences (Potochnik 2013). Scientists adhering to different, incompatible approaches might appear to constitute a fundamental theoretical divide, and the involved scientists may even construe it as such. But the view set out in this section enables an alternative interpretation. Distinct scientific products, incorporating different idealizations, are helpful in the pursuit of different aims. Each approach may be successful given its specific aim, and no approach succeeds in addressing all the relevant aims. This different interpretation transforms what may otherwise appear to be a crisis or at least limitation of an investigation into an expected feature of successful scientific research. Even if scientists do not emphasize the different concerns that guide their work or mistakenly interpret different aims as disagreements about findings, differences among specific aims can often account for and explain away apparent disagreement.

To summarize, in this section, I have motivated the idea that the different aims of science and even specific deployments of individual aims are best pursued separately and in fact are pursued separately. Different scientific products are best suited to different aims of understanding, prediction, guidance of policy or action, and so on. This is due in part to the variety of purposes served by idealizations and so the variety of idealizations incorporated into models and other representations. It is also deeply related to conceiving of science's epistemic aim as not truth but understanding. If scientific results were fully true, they may also be the best way to generate explanations, predictions, policy, and all the rest. But if science's epistemic success consists of understanding, then this is just one among several aims, aims that I have suggested must be pursued independently. I have suggested that this outcome is ultimately due to the limited powers of human cognition and action, when faced with the exceedingly complex phenomena under scientific investigation.

In light of this variety of scientific aims and multiple deployments of individual aims, one might wonder what determines the aim served by any individual scientific project. If there are many different, potentially applicable scientific aims, by what standard should we judge the success of a given scientific product? Some generalizations can be made about what determines the applicable aims. First, as outlined above, the particular research focus directly influences the features of a phenomenon that contribute to understanding. The research program also influences the broader determination of the importance of prediction, explanation, guidance of action, and so on. Consider yet again the scientific projects discussed in Chapter 3. Some research agendas in behavioral ecology focus on the role of natural selection in producing a behavior, whereas research on epigenetics will foreground environmental influence on genetic factors. For both of these focuses, prediction is much less central than it is in climate science. Another influence on what aim is applicable are the features of the practitioners of science and their target audience. A primary aim of climate research is to furnish us with predictions, and particular types of predictions, in part because climate change is expected to have vast economic, political, and other consequences for humans. We want to know what to expect and how we can alter the trajectory. In turn, whether a prediction is intended for other researchers or for policymakers may lead to an emphasis on, respectively, maximizing accuracy or minimizing risk.

These considerations shape specific research aims; there is also a question of how to suss out the aim by which a particular scientific product should be judged. Often this can be discerned by charitable interpretation of the research conducted, including its setup, the conclusions drawn, and its research context and social context. In some cases, such interpretive work may not eliminate all uncertainty regarding what aim is pursued and thus by what standard the work should be judged. Indeed, sometimes there may be a true ambiguity in what the research is aimed to accomplish. It's also common for there to be a mismatch between the pursued aim and conclusions drawn from the research by other researchers or, especially, by the popular media. A first step toward eliminating such confusions is for scientists and consumers of science alike to recognize that any given scientific product is in service of some particular aim and may well not contribute to any other scientific aims.

4.2 Understanding, Truth, and Knowledge

Now that I have developed my account of science's aims and the relationship among them, I return again to the central idea that science's epistemic aim is understanding. Some features of that idea require more discussion. I have suggested that understanding can count as an epistemic achievement in its own right, even as it does not require truth. On this, I part ways with many accounts of understanding. Accordingly, I should say more about what I take the nature of understanding to be. After doing so, I then consider the ways in which, on this view, truth and knowledge may still factor in to science. Taking understanding to be the epistemic aim of science necessitates a reconsideration of the role of truth and the nature of scientific knowledge, but both may retain a place in a full account of science's success. Finally, I conclude this chapter with a brief discussion of some of the positive implications of taking science's epistemic aim to be understanding. Recall that my project in this book is intended to buttress, not undermine, the success of science. The main thesis of this chapter in particular demonstrates how scientific products reflect their human creators in legitimate ways, and it motivates a promising interpretation of scientific progress and the present state of our scientific understanding.

4.2.1 THE NATURE OF SCIENTIFIC UNDERSTANDING

Positing understanding as the epistemic aim of science invites the question of what scientific understanding is. I have already introduced some features that, in my view, a concept of scientific understanding must have. Understanding, as a cognitive achievement, has a dual nature. On one hand, because understanding is a cognitive state, its possession depends on features of the cognizer. This is significant for my purposes because it means that, unlike truth, what promotes understanding is partly determined by features of the practitioners and consumers of science. This enables deviations from truth to be justified with reference to the limitations and particularities of scientists and their audiences. Understanding is, accordingly, not factive. On the other hand, understanding is also an epistemic achievement. The subjective experience of a sense of understanding is not sufficient for possessing understanding. Characterized very abstractly, genuine understanding requires not only the proper cognitive state but also the proper relationship between that cognitive state and the world. This is what gives understanding epistemic significance, and it's what the requirement of epistemic acceptability is supposed to ensure. Here I say more about these features of understanding, drawing from research in cognitive psychology, epistemology, and philosophy of science. This enables me to characterize the concept of understanding that is relevant for my purposes and to articulate the features of *scientific* understanding in particular.

Empirical research in cognitive psychology has yielded insight into some of the cognitive characteristics that promote human understanding. Much of the psychology research on understanding explicitly addresses explanation. I follow a large number of philosophers and psychologists in positing a close relationship between explanation and understanding, and so I draw conclusions about understanding partly on the basis of research into the cognitive aspects of explanation.[5] Alison Gopnik (1998) argues that explanation is a mark of what she calls the "theory-formation system," which aims to identify causal relations. This is not supposed to be special to scientific explanation. Gopnik suggests it is a much broader cognitive phenomenon in humans' everyday reasoning, although most apparent in children and scientists. She emphasizes the role played by active intervention on the part of the cognizer, and she conceives of explanation as a phenomenological consequence of forming or adapting theories in response to such interventions. As Tania Lombrozo (2011) surveys, empirical research suggests that the act of explaining plays certain specific cognitive functions. These include improvements in learning about both general patterns and causal structure but not improvements in learning other types of content. Lombrozo also discusses research demonstrating that people show a preference for simple explanations and explanations that are general, that is, that apply to a large number of phenomena. Lombrozo and Carey (2006) investigate the conditions in which teleological, or functional, explanations are judged to be acceptable. They conclude that the acceptability of a functional explanation requires that the function has acted as a causal influence and also that this causal influence conforms to a general pattern.

This research suggests that understanding is promoted by simple generalizations about broad patterns, especially regarding causal relationships, and perhaps especially causal relationships grounded in interventions, or manipulations. Each of these attributes is prima facie cognitively useful. Simple representations are easier to grasp than are complicated representations, and both broad patterns and causal information help in reasoning about possible future occurrences, including outcomes of future interventions. All of this corroborates my suggestion that scientific understanding is produced by the identification of causal patterns. The relevant concept of causation is manipulationist, which I have suggested is epistemically basic, grounded in humans' actual and possible interventions on our world. As Gopnik (1998) emphasizes, this is how children and scientists conduct investigations on their world. This gives rise to the two sources of human understanding: patterns and causes.

Williams et al. (2013) provide additional empirical support for the idea that human understanding is furthered by generalizations about broad patterns, and they show that this can lead to a propensity to overgeneralize. These authors cite "the remarkable human capacity to discover patterns and construct generalizations from sparse observations" (1011). The risk of overgeneralization and of ignoring exceptions is undoubtedly a cost: it inhibits certain types of learning, especially about the specificities of individual phenomena. And yet, Williams et al. point out in the concluding paragraph of their paper that "ignoring exceptions may sometimes help learning. For example, when there is substantial variability in observations, exceptions could erroneously lead learners away from noisy but reliable patterns" (1013). For scientific observations, this kind of variability is a straightforward consequence of causal complexity. Accordingly, in my view, the benefit of ignoring exceptions that Williams et al. mention is deeply related to our science's focus on patterns at the expense of accuracy. The cognitive benefit of simple generalizations comes in conflict with the causal complexity of our world. When it does, the aim of scientific understanding can motivate departures from accuracy—even, as we have seen, quite significant departures.

This cognitive science research gives some insight into the nature of understanding and why it is useful to humans, and it also accords with some of what philosophers have had to say about understanding. I mentioned earlier in the chapter that Grimm (2006, 2011, 2012) and Strevens (2013) take the psychological state of understanding to consist of a kind of grasping. Grimm articulates this in manipulationist terms: "to grasp how the different aspects of a system depend upon one another is to be able to anticipate how changes in one part of the system will lead (or fail to lead) to changes in another part" (89). Grasping in this sense is, then, a kind of ability. In particular, it is an ability to predict the effects of certain kinds of in-principle interventions. The research surveyed just above suggests that this ability is improved by an emphasis on broad, simple patterns and, accordingly, the limited focus this requires. It also can lead to inaccuracies that arise from ignoring deviations from and exceptions to a general pattern. For this reason, and as I argued earlier in the chapter, *scientific* understanding at least does not require belief. On this, I part ways from Grimm, who takes understanding to involve not simply grasping but also belief and to be a subspecies of knowledge (see especially Grimm 2006).

On the other hand, genuine understanding involves more than just occupying a particular psychological state. J. D. Trout (2002, 2007) argues that a felt sense of understanding is not a sign that genuine understanding has been achieved. He puts the point in terms of truth: he argues that a sense of understanding is produced by biases that actually diminish truth-tracking ability, including overconfidence bias and hindsight bias. Overconfidence bias is the human tendency to have more confidence in our beliefs than their accuracy warrants, while hindsight bias is the human tendency to consider an unexpected outcome or finding to be, in hindsight, expectable. Trout is partly right but partly wrong. He is right that, if understanding is to play an epistemic role, it cannot be merely a psychological state. But Trout puts the point too strongly when he says that our sense of understanding is *due* to biases like overconfidence and hindsight. Both of these biases can lead a felt sense of understanding to diverge from genuine understanding, but this does not mean they are what generates the sense of understanding in the first place. I have suggested, with some basis in empirical and philosophical research, that a sense of understanding is instead generated by apparently grasping causal patterns (whether accurately or erroneously). Trout also claims that a sense of understanding cannot play a justificatory role, but this is also too strong. A felt sense of understanding cannot be the basis for the assertion that this is genuine understanding, an epistemic achievement; surely this is what Trout means. Yet a felt sense of understanding is, in another way, an essential part of diagnosing whether there is genuine understanding. The other half of understanding's dual nature is also relevant: you cannot understand without apparently grasping, without being in the psychological state of understanding. Scientific understanding is not produced merely by information about the world; it is produced when information about the world is of the right sort to induce in us a felt sense of understanding.

How, then, can we tell if understanding is actual and not merely apparent? For this, the causal pattern apparently grasped must be *real*. In Chapter 2, I motivated the idea that causal patterns are real, drawing inspiration from Dennett (1991). Briefly, for a causal pattern to be real, it must be embodied (to some degree or other) in some range of phenomena. When all posits contributing to a representation satisfy the requirement of epistemic acceptability I outlined above, this should ensure the represented causal pattern is real. Importantly, what is required for epistemic acceptability varies with research interests, that is, with the epistemic purpose to which a representation is put. This reflects how, in order to generate understanding, a grasped causal pattern must also connect up in the right way with the pertaining research interests. Phenomena embody lots of causal patterns; grasping any old causal pattern embodied by some phenomenon won't lead to an understanding of that phenomenon. The grasped causal pattern must relate in the right way to the inquiry for it to produce understanding and for it to produce the felt sense of understanding that accompanies this epistemic achievement.

As an illustration of distinguishing real from apparent understanding, consider again the contrast between idealized scientific representations and astrology. Like Wimsatt (2007), Elgin (2007) contrasts the sort of falsity involved in these cases. Regarding astrology, she says, "A coherent body of predominantly false and unfounded beliefs does not constitute an understanding of the phenomena they purportedly bear on. So despite its coherence, astrology affords no understanding of the cosmic order" (35). It does not matter how much of a felt sense of understanding a horoscope generates in a believer of astrological causes. Empirical research has failed to confirm any such causal relationships. However, on my account, it is not the falsity per se that is problematic but the false elements' intended representational role and how the resulting representation relates to research interests. Astrology does not yield actual understanding because its false and unfounded posits include those that are intended to accurately capture causal influence. For this reason, astrology's posits are not epistemically acceptable. And for this reason, astrology represents causal patterns that do not exist.

In my view, then, scientific understanding consists of grasping a causal pattern, and it is distinct from belief and also—for at least that reason—distinct from knowledge. But I follow Grimm in taking grasping to be a form of epistemic success. Understanding is not merely a subjective state but the epistemic accomplishment that produces such a state. For me, this entails that all posits contributing to understanding be epistemically acceptable, which ensures the causal pattern responsible for the understanding is real. Some of this might be distinctive of scientific understanding in particular. Perhaps scientific understanding is grounded more in manipulability than are all forms of everyday understanding. Perhaps the extent to which deviations from truth are conducive to understanding in science is not matched outside of science. Because science is a social enterprise, the aim of scientific understanding motivates products for cognitive consumption by different audiences, products that can be variously engaged with and reapplied. Perhaps some paths to everyday understanding are less thoroughly and crucially social.

4.2.2 THE ROLE OF TRUTH AND SCIENTIFIC KNOWLEDGE

The view I have developed in this chapter has rather dramatic implications. It unseats truth as the epistemic aim of science and suggests that many of the products of science are not things we believe to be true or even approximately true. There is still an epistemic standard governing the successful products of scientific research: to contribute to legitimate understanding, any posit must satisfy the requirement of epistemic acceptability I have articulated. But epistemic acceptability is relative to the features of scientists and their specific projects in a way that truth is not, and it does not amount to requiring approximate truth. The clearest illustrations of this are idealizations themselves, which are quite far from the truth but, in the right circumstances, are epistemically acceptable nonetheless. So, in my view, science simply is not after the truth. There are some important ways in which truth still may be involved in the scientific enterprise, but in each case, it is only a means to other ends.

Achieving genuine understanding requires *some* type and degree of accuracy in virtue of the requirement of epistemic acceptability, but this requirement is weaker and variable in a way that truth and even approximate truth are not. As I have emphasized, the threshold for what is epistemically acceptable varies according to the role a posit plays in a representation and the epistemic purpose to which that representation is put. An idealization is acknowledged to be false, so can diverge radically from the truth so long as that divergence contributes to achieving understanding in light of the specific research interests. Epistemic acceptability is more closely aligned with truth for some other types of posits. Truth may also more uniformly contribute to other aims. I have argued that science has other aims aside from the epistemic aim of understanding, and some of those might best be served by true claims. For example, perhaps accurate predictions are best described as true assertions about future states of affairs. But notice that truth is also only relevant in this circumstance insofar as it serves a different scientific aim, namely, prediction. Furthermore, as the case study of climate science in Chapter 3 illustrates, even if our best predictions are true assertions, these are often produced by highly idealized representations. Predictions, like understanding, are generated with the use of convenient untruths.

So truth or accuracy in some respect and to some degree is required for genuine understanding, and truth might more uniformly contribute to other scientific aims, like prediction. But, one might press, is truth itself not also an epistemic aim of science, at least sometimes? Consider three possibilities. First, perhaps sometimes straightforwardly true claims are the best way to generate understanding. Examples of this might include claims like that Saturn has a moon larger than the planet Mercury, or that life on Earth has evolved from one or a few common ancestors. In such cases, it is impossible to distinguish the aim of truth from the aim of understanding, for the same scientific products accomplish both. These are cases in which departures from truth do not further understanding. I suspect this circumstance is limited to particular claims about individual states of affairs, like the brief examples above. All other scientific representations regularly depart from truth by employing idealizations in the way surveyed in Chapter 2. Second, I have said that general patterns are valuable for human understanding, but perhaps sometimes scientists value more accurate representations—representations that are closer to the truth—over representations of general patterns. This must be right: sometimes exceptions or individual instances are central to research interests. But this is to be expected if understanding is the epistemic aim, for I have stressed that the nature of the understanding sought is paramount to what contributes to understanding. When research interests are narrow or specific, understanding may consist of grasping a more nuanced causal pattern. As with the first scenario, this possibility is to be expected if the epistemic aim of science is understanding.

Third and finally, consider the possibility that scientific products with the epistemic aim of truth might be developed at the expense of their contribution to understanding. Unlike the first two, this possibility pits the supposed aim of truth against the aim of understanding so they can be prized apart. I suspect the pursuit of understanding always carries the day. I simply can't imagine scientists developing products that produce *less* understanding of a phenomenon of interest simply to be more veridical—that is, unless there is some other aim in play, such as prediction. I can't provide a general argument for this conclusion since I take there to be many aims of science and possibly an open-ended and evolving set of aims. It's always possible that truth simpliciter emerges as one. But I expect that, for cases that may seem to fit this description, closer investigation of what they are taken to accomplish reveals that they are after understanding, prediction, or another aim beyond simple truth after all. This expectation might be due to my impoverished imagination, and I am happy to entertain possible counterexamples. But I would wager that, when the world's complexity and variability inhibit representations that are both accurate and comprehendible, comprehension (or predictability, action in a short time span, etc.) always trumps simple accuracy.

The idea that science does not aim at truth, but rather understanding, does not entail antirealism. In its most abstract formulation, that debate regards whether our best science has achieved epistemic success or whether science aims to achieve epistemic success (Chakravartty 2013). I have not challenged the idea that there is an epistemic aim to science; my criticisms are of the idea that this aim is best articulated as truth. My view draws into question any version of realism that defines epistemic success as truth or even approximate truth, but other articulations of realism are readily available. Indeed, that I have posited an epistemic aim for science is opposed to purely instrumentalist versions of antirealism. Science is sometimes after accurate predictions, but many of its activities resist that interpretation. Any realist or antirealist construal of science must account for its diverse epistemic and nonepistemic aims. The position developed in this chapter constrains what form a realism or an antirealism might take, but it does not pretend to settle that issue.

A related issue is whether the aim of understanding, as I have articulated it, allows for the production of scientific knowledge. There is something of a puzzle here. Jettisoning the idea that understanding requires true beliefs entails that it is not a species of knowledge, for knowledge is generally taken to be factive. That is, only facts, or truths, can be known. Here is where it is important that causal patterns are real. If causal patterns are real, then scientific knowledge consists of truths about causal patterns. Grasping those truths about causal patterns comprises understanding of the phenomena embodying the patterns. As we have seen, truths of causal patterns are by and large partial untruths about phenomena, accomplished with the use of idealizations. The resulting knowledge is thus not properly construed as knowledge of those phenomena. There is a connection between knowledge and understanding, but in an attenuated form. Scientific knowledge of causal patterns is the route to scientific understanding of phenomena.

So, if causal patterns are real, then idealized representations can truly depict causal patterns, even as they inaccurately depict actual phenomena. Our scientific representations generate understanding of phenomena in virtue of being true of causal patterns. Here, then, is another role for truth in science's epistemic achievements. But this role too is in service of human understanding. Truths about causal patterns are sought because grasping them generates human understanding—or else they contribute to some other human-centric aim, like prediction or quick action. Causal patterns are of interest only when they contribute to answering questions humans have posed about the world we encounter. Recall that any given phenomenon embodies many causal patterns, probably infinitely many. All of those patterns are real. But which of these many patterns scientists seek knowledge of depends on the nature of their interest in the phenomenon. Scientists' cognitive and practical goals, goals related to the tangible world of objects and events, drive their interest in causal patterns. Ultimately, when it comes to science's epistemic aim, knowledge and truth provide variable-length tethers to the material world, tethers that are in service of human understanding.

I conclude this chapter by considering a few of the virtues of an account of science as focused ultimately on understanding rather than truth or knowledge. First, this achieves what I set out to do in this chapter, namely, account for rampant and unchecked idealization by showing how idealization can contribute directly to science's aims, including its epistemic aim. Idealizations are always at the expense of truth, but they can contribute directly to understanding. This is possible because understanding has a cognitive dimension. The cognitive aspect of understanding provides a legitimate way in which science's epistemic achievements reflect features of scientists and their audiences, in general qua humans and also their particular characteristics and concerns. I develop this idea in greater depth in Chapter 7.

A related implication of the cognitive aspect of understanding regards the status of what have been called epistemic values. Epistemic values are usually taken to be justified by their contribution to truth tracking (if they are justified at all). My view enables a somewhat different interpretation. Epistemic values may qualify as epistemic not by contributing to truth tracking but by contributing to the cognitive dimension of understanding. Simplicity and generality, for example, have proven difficult to justify via their contribution to truth.[6] But even if they don't increase the likelihood of truth about the world we investigate, both are features of scientific representations that we humans find illuminating. Psychology research discussed in the previous section suggests that humans prefer explanations that are simple and general (Lombrozo 2011; Williams et al. 2013). Because understanding is both a cognitive state and an epistemic achievement, epistemic values include properties that contribute to either dimension. Note also that whether a property contributes to the cognitive dimension of understanding may vary among individuals, depending on their research interests or other differences. However, I don't expect this to be the case for simplicity or generality. Some philosophers have questioned whether there is a well-founded distinction between epistemic and nonepistemic values. What I have said here does not undermine the distinction, but it does limit its significance. Like nonepistemic values, epistemic values might only reflect our characteristics or preferences, and some of those values might vary among individuals.

Taking the epistemic aim of science to be understanding also inspires a rather cheery interpretation of scientific progress. If the epistemic aim of science is truth, then science has fallen short of this aim again and again. We face the problem of the pessimistic meta-induction: why think our current theories are true when their many predecessors were not? If, in contrast, the epistemic aim all along has been understanding, then the outlook is much rosier. Many superseded scientific theories posit radically different views of the world. They thus must have been radically wrong. The Earth cannot be both the stationary center of the universe and also revolve around the Sun. But taking these theories to represent causal patterns enables a subtler accounting of what they had right. Of course, some scientific conjectures are simply wrong: they posit causal patterns that do not exist. But many previously accepted theories latched on to causal patterns that *are* embodied in phenomena, even if they were later replaced by other theories that depicted other patterns, including sometimes more refined relatives of the original pattern. The closed orbit theory of ionization discussed in Chapter 3 posits classical electron orbits because these capture something crucial about ionization patterns, even though they'd been rejected as false. Indeed, Bokulich (2008) discusses a number of ways in which classical mechanics lives on, despite the theory's rejection. My interpretation of science's epistemic aim enables this and many other episodes of scientific change to be interpreted in a cumulative way.

In the same way, we can be confident of our present science's epistemic success. Our current best theories are probably not true. Some of our current best theories even contradict one another. All of those theories, along with many other kinds of scientific products, stand a good chance of being jettisoned or significantly altered sometime in the future. But even if that comes to pass, it does not undermine their contribution to human understanding. If humans now use these products to grasp real causal patterns, then it does not matter if they aren't true. Science is always in revision, but it is amassing ever more understanding.

5

Causal Pattern Explanations

Chapter 4's emphasis on understanding leads naturally to a discussion of scientific explanation. Explanations have classically been taken to be the means for creating understanding, although there is a growing debate over whether scientific understanding can be produced without recourse to scientific explanations. The topic of scientific explanation also connects with this book's focus on idealization, as there is a well-developed literature on how idealizations might contribute to scientific explanations. I suspect this is because a focus on explanation is thought to create a role for consideration of humans' cognitive requirements, and it is in this human-centric regard that idealizations seem to make a clear contribution to science. Consideration of human cognitive requirements also figures into other emphases in philosophical treatments of scientific explanation. For instance, Carl Hempel (1965) motivated the classic deductive-nomological account of explanation by pointing out that this style of explanation shows that "the occurrence of the phenomenon *was to be expected*." He continues, "It is in this sense that the explanation enables us to *understand why* the phenomenon occurred" (337). Hempel invokes a specific human cognitive state—the recognition of expectability—as the reason explanations generate understanding. Another example of considerations of human cognition appearing in treatments of scientific explanation is the idea that explanations are improved by being general (Putnam 1975; Garfinkel 1981; Jackson and Pettit 1992). This idea is usually motivated by appealing to what general explanations make salient, or cognitively accessible, to their audiences.

Yet few philosophical accounts of scientific explanation explicitly invoke considerations of human cognitive requirements. Instead, most accounts exclusively emphasize the relationship that must obtain between explanations and the world. Debates have raged over whether explanations should feature laws of nature, the unification of known facts, causal information, and so on. But the connection between explanation and understanding, together with the ideas introduced in Chapter 4, demonstrates the centrality of human cognitive requirements to the production of understanding and thus to scientific explanation. Not only must explanations connect to the world in the proper way, but they must also connect to their human audience—the producers and consumers of explanations—in the proper way. I develop this idea in section 5.1. I then show how the resulting view situates explanations at the center of science, as the primary vehicles of scientific understanding. In light of that connection, the positions I have developed regarding the nature of understanding and the somewhat distant relationship between understanding and truth also provide insight into the nature of scientific explanation. In section 5.2, I motivate an approach to explanation on the basis of these insights. There are three defining features of that approach. Causal patterns' central role in securing scientific understanding places them also at the heart of our explanations. Yet causal complexity, along with rampant and unchecked idealization, necessitates an approach to explanation according to which an explanation produces, at best, understanding of only certain features of a phenomenon. This elevates contextual considerations to a central position in this account of explanation, for which features are to be understood depends on the audience's immediate interests. And, finally, the resulting view creates the need for a minimal standard of adequacy that every explanation must satisfy but a standard that still allows ample room for persistent idealization.

5.1 Explanation, Communication, and Understanding

My task in this section is not to advocate for any particular account of explanation. Instead, I consider the prior question of what an account of explanation should accomplish. Strevens (2008b) and Craver (2014) both explicitly defend the view that the task for an account of explanation is to indicate what facts out in the world explain any given event. Craver says, "The philosophical dispute about explanation, from this ontic perspective, is about which kinds of ontic structure properly count as explanatory and which do not" (29). In contrast, my emphasis on how human cognitive needs shape scientific understanding leads instead to a focus on explanations as particular acts of communication. On this alternative approach, an account of explanation should indicate what representation counts as a satisfactory answer to any given request for explanation. This has been described as a communicative, versus ontic or ontological, approach to explanation. After motivating that approach, I move on to another question about explanation distinct from the task of providing an account of explanation. Namely, I consider what role explaining plays in the scientific enterprise. In Chapter 4, I defended the idea that science has many distinct and competing aims; here I ask where explanation fits in that schema. My conclusion is that, because explanations furnish our scientific understanding, they are at the heart of science. Explanations are crucial to science's epistemic aim. This accords with the central role that both scientists and philosophers typically ascribe to explanations. And yet, in my view, it is the communicative sense of explanation that occupies this role. So let's start there.

Several philosophers have pointed out that the term "explanation" is used to mean different things. According to Carl Craver (2014), for example, the subject that takes the verb "to explain" might be four different types of things: something out in the world, a person, a scientific representation, or a mental representation. Imagine a teaspoon of table salt settling at the bottom of a beaker of water rather than dissolving. One sense of the verb, what Craver calls the ontic use, enables us to say things like "The solution's having reached its saturation point explains why no more salt would dissolve in it." In this case, something out in the world, a state of affairs, is doing the explaining. In a second sense, the communicative use, we say things like, "The chemist explained to her audience why no more salt would dissolve in the solution." And in a third, representational use, we could say, "The solubility graph explains why no more salt would dissolve in the solution."

Craver (2014) argues that the ontic sense of explanation is basic. He says,

> Scientific explanations are constructed and communicated by limited cognitive agents with particular pragmatic orientations. These topics are interesting, but they are downstream from discussions of what counts as an explanation for something else. Our abstract and idealized representations count as conveying explanatory information in virtue of the fact that they represent certain kinds of ontic structures (and not others). (29)

Michael Strevens (2008b) agrees. In his view, "what explains a given phenomenon is a set of causal facts.... The communicative acts that we call explanations are attempts to convey some part of this explanatory causal information" (6). As Strevens notes, this ontological focus is traditional for philosophical accounts of scientific explanation.[1] He provides a vivid metaphor: "a philosopher of explanation will ... occasionally discuss communicative conventions, just as an astronomer might study atmospheric distortion so as to more clearly see the stars" (6). On an ontic approach to explanation, then, communicative requirements are taken to merely distort or edit ontic explanations, and the latter is the appropriate target for philosophical accounts of explanation. Just as atmospheric distortion can only influence our view of the stars, not the stars themselves, so too are ontic explanations uninfluenced by communicative requirements.

Craver (2014) is right to say that representations count as explanatory in virtue of their relationship to the world, to what he calls "certain kinds of ontic structures." Scientific explanations must be connected in the proper way to features of the world; this is what allows them to convey information about that world and information of the right kind to be explanatory. Put most broadly, an explanation must reflect what is (in some sense) responsible for the event to be explained. This means that explanations must depict dependence relations: what, out in the world, bears responsibility for the phenomenon's occurrence. And Strevens (2008b) is right that the relationship scientific explanations must bear to the world has received the lion's share of philosophical attention. The kind of responsibility that is explanatory is primarily what is at issue among different accounts of explanation. Thus, the deductive-nomological approach posits nomic responsibility as explanatory, the causal approach posits causal responsibility, the mechanistic approach posits causal interactions among hierarchically organized entities, and so on. Strevens additionally distinguishes between the determination of the right form of responsibility—on his view, causal facts are explanatory—and the determination of which of the facts about responsibility are explanatorily relevant for a given explanandum. This results in what he calls a two-factor account of explanation, where the first factor is a metaphysical account of causal relevance and the second factor is a separate account that determines which causal facts are explanatorily relevant. But even in this two-factor account, both factors regard the relationship explanations must bear to the world.

However, bearing the proper relationship to the world is only one of the tasks at which explanations must succeed. They also must establish a connection of the proper sort to human cognizers, to those seeking an explanation. The ontological features of an explanation may seem to have priority, and I do not dispute that it is necessary for an explanation to "hook up" to the world in the right way. But it is just as necessary for an explanation to "hook up" in the right way to those seeking an explanation. There is no explanation unless something is (at least potentially) explained, and the latter is subject not only to facts about the world but also to facts about cognition. Shifting the focus to the relationship between an explanation and its audience foregrounds the sense of explanation as a communicative act. Facts out in the world do not in themselves bear the proper relationship to human cognizers that is needed for explanation. Those facts must be represented and communicated—and in the right way—in order for that connection to be forged. These features of explanations are thus also important to explanatory success.

As one might expect in light of the attention the ontological features of explanation have received, philosophical discussions of explanation have tended to downplay the significance of scientific explanations considered as communicative acts. A number of influences on the actual explanations formulated in science have traditionally been relegated to the category of the "pragmatics" of explanation. This terminology suggests a parallel with linguistics, where pragmatics is the study of particular speech acts and variation in meaning due to context. The influences traditionally included in the category of pragmatics of explanation are the particular features of an explainer and an explanation's audience, including their cognitive features, epistemic circumstances, interests and concerns, and the features of scientists and humans in general, including our cognitive features, epistemic circumstances, and shared interests and concerns. In short, everything relevant to an explanation's relationship to its audience is typically deemed merely pragmatics. It is sometimes explicitly acknowledged that what explanation is in fact generated depends in part on such pragmatic considerations, but philosophical accounts of explanation by and large follow the prioritization that Craver (2014) describes: first one must determine what out in the world counts as an explanation, then one might choose to consider the "downstream" questions regarding pragmatics and communication. In actuality, few philosophers find reason to turn to these questions deemed secondary. Indeed, in conversation, Craver has referred to these questions as belonging in "the dustbin of pragmatics." On some views, the pragmatics of explanation is nothing special, that is, is in no way distinct from the pragmatics of linguistic communication more generally (Lewis 1986).

There are exceptions to this approach of downplaying the relationship explanations must bear to their audience, if only a few. Silvain Bromberger (1966) suggested that explanations should be taken to be answers to particular why-questions. Bas van Fraassen's (1980) pragmatic account of explanation also emphasizes the primacy of the audience's concerns in shaping explanations. Yet van Fraassen suggests that the audience's influence on explanation consists in determining what type of responsibility relation out in the world is explanatory; so despite his emphasis on the audience, his account of explanation is still primarily framed as an account of ontological explanatory relevance. Peter Achinstein's (1983) approach to explanation begins with the act of providing an explanation. For Achinstein, whether something counts as a good explanation depends on both the explainer and the audience for the explanation. Finally, much more recently, Cory Wright (2012) argues that the default sense of explanation is communicative in nature.

I agree with Wright that explanation is at root an act of communication, and I agree with Achinstein that what counts as a good explanation depends on both the explainer and the audience. Sidelining the communicative purposes to which explanations are put is a mistake. Bearing the proper relationship to an audience is a requirement on any explanation, and satisfying that requirement critically shapes the nature of scientific explanations. The basic reason for this is by this point familiar: the realities of a science accomplished by humans in a complex world. Scientific explanations must be comprehendible by humans—they must generate human understanding. And yet, in our causally complex world, there are many—possibly countlessly many—factors upon which any given event depends. This is so whether the explanatory dependence relations are taken to be causal, causal-mechanical, law-like, or some other form(s) of dependence. An account of explanatory relevance that does not explicitly consider the relationship between explanations and explainers leaves unanswered the questions of which of many dependencies should be featured in any given explanation and how those dependencies should be represented.

Those who pursue an ontic approach to explanation might agree with what I've said so far, at least in principle. Lewis (1986) acknowledges that what is represented in an explanation and how it is represented are both influenced by the audience, but he takes this to be so only for explanations actually formulated at a given point in time, not to what he calls "*the* explanation." According to Lewis,

> Among the true propositions about the causal history of an event, one is maximal in strength. It is the whole truth on the subject—the biggest chunk of explanatory information that is free of error. We might call this the *whole* explanation of the explanandum event, or simply *the* explanation. (1986, 218–219)[2]

For Lewis, a philosophical account of explanation regards this, *the* explanation—what we might think of as the ontic explanation. This move evokes Strevens's comparison of communicative conventions for explanation to atmospheric distortion for the stars. Both consider the only distinctive question about scientific explanations to be the explanatory dependence relations, and both take that question to be beyond the reach of human influence.

Yet I take explanations' communicative features to be more central than this allows. An explanation's required relationship to its audience even influences the relationship that explanation must bear to the world, that is, its so-called ontological features. I have suggested that an explanation's audience influences what should be represented and how it should be represented to generate a satisfactory explanation in any individual instance of explaining. One element of this influence is due to the audience's specific interests, that is, what they seek to understand. The combination of what and how an explanation represents, together with what the audience seeks to understand, determines what dependencies are featured in an explanation. This point will be corroborated by the particular approach to explanation I develop in section 5.2; see also Potochnik (2016). An explanation's audience thus helps determine the nature of the explanatory facts, that is, the ontic explanation. A maximal explanation like Lewis proposes does not eliminate the audience's influence. Explanations don't only err by omitting information they should include; by including information they should omit, explanations indicate dependencies that do not obtain. A maximally inclusive explanation would feature too much information to succeed with any audience, not just by violating communicative conventions but also by signaling dependencies that do not obtain. Ontic explanations are indelibly shaped by the communicative purposes of explanations in general, as well as by the particular scientific context in which an explanation is formulated—that is, the nature of the why-question and the nature and circumstances of those who formulate that question.

Consideration of an explanation's audience, or those seeking the explanation, is thus essential to providing an account of explanation. An explanation's communicative purposes are not downstream but *upstream* from considerations of what "ontic structures" our explanations should represent. By this I mean that the audience for an explanation, the explanation's communicative context, influences both representational and ontological features of explanations. They influence what and how an explanation should represent, and this in turn results in the explanation featuring different dependence relations. To be clear, this does not entail that the *type* of explanatory dependence relation is determined by scientists' interests. Rather, the point is that, whatever type(s) those are—causal, nomic, mathematical, or other—which particular dependence relation explains some phenomenon depends on the specific interests behind the request for explanation. To determine what explains some phenomenon, then, one must first ascertain the research focus that occasions the explanation. Only then can one pose the question of what out in the world explains. It's possible that traditional philosophical accounts of explanation assume this matter of the research agenda has been resolved in any given instance of explaining, before an account of explanation focused on the question of ontological dependence gets going. But this does not render communicative context unimportant; it simply makes it invisible when it is actually primary.

The communicative purposes to which explanations are put—that is, their relationships to human explainers in the abstract and in particular instances—help account for other features of explanations as well. To begin with, this accounts for what is distinctive about explanation as a scientific aim. Several philosophers have justified the value of scientific explanations with the claim that explanatory information is what would be missing for Laplace's demon, a creature possessing all information about the current state of the universe and the (presumed deterministic) laws of nature and, on that basis, capable of predicting all future states and retrodicting all past states. Salmon (1978) uses this strategy to help motivate a causal-mechanical approach to explanation, Strevens (2008b) to motivate the view that all higher-order causal relations are instead explanatory relations, and Douglas (2009a) to motivate the use of explanations for prediction. All three reference the role of explanation in producing understanding.[3] From Douglas (2009b):

> The value of explanations can be rescued ... when we recall that we are not Laplacian demons. We do not have perfect predictive capacities. And even if we had a factually complete account of this very complex world, and perfect predictive capacities to utilize those facts, the complexity would be overwhelming to us. We are finite beings, with finite mental capacities. We need explanations to grapple with all of this complexity. Explanations help us to organize the complex world we encounter, making it cognitively manageable (which may be why they also give us a sense of understanding). (54)

Citing explanation's usefulness for limited human cognizers implicitly directs our attention to its communicative purposes. It is explanation in a communicative sense, explanations actually formulated for specific human audiences, that is relevant here.

The communicative purposes of explanations are also relevant to the connections that have been posited between explanation and idealization. Those who defend the scientific value of idealizations largely base that defense on arguments for how idealizations can contribute to scientific explanation. Instances of this strategy are so numerous that I leave it to readers to generate their own examples. I suspect that the communicative roles of explanation, as well as the resultant relativity of good explanations to the features of the individuals who develop them, is the reason that this strategy is so widely adopted. Focusing on explanation warrants an emphasis on how the practice of idealization is cognitively useful to us, that is, how it facilitates our understanding. But this is so only if explanations are shaped not only by the relationship they must have to the world but also by the relationship they must have to those who seek explanation. Accordingly, positing an explanatory role for idealizations has consequences for how an account of explanation should be approached. This too indicates that the communicative sense of explanation must be given priority. Moreover, in earlier chapters, I motivated the idea that what idealizations are appropriate depends not just on the world, such as the causal facts, but also on facts about scientists themselves, including their particular research interests. If this is right, then any philosophical account of explanation that sanctions idealized explanations must explicitly consider the relationship that explanations bear to human explainers.

One final significance of the communicative purposes of explanation I will discuss in greater depth. This is its role in relating explanation to understanding. As I touched on above, scientific explanations have classically been taken to be the means for generating scientific understanding. Many philosophical treatments of explanation assert such a connection between explanation and understanding. Some philosophers have also explicitly defended understanding as the goal of explanation (e.g., Grimm 2010) and explanation as the only method of achieving understanding (e.g., Strevens 2013). But this is at odds with the nearly exclusive philosophical focus on the relationship that explanations should bear to the world and resulting neglect of the relationship between explanations and explainers. Consider that Hempel (1965) holds both that explanations show us that a phenomenon was to be expected, thereby enabling us to understand the phenomenon, and also that explanations require demonstrating via logical deduction how a phenomenon depends on a law of nature. But nothing guarantees that the former is accomplished, and uniquely accomplished, by such derivations. Indeed, it is now widely held that this is not so. Strevens (2008b) says that "[he takes] scientific understanding to be that state produced, and only produced, by grasping a true explanation" (3). As we have seen, he also specifies that his account is exclusively about the ontological sense of explanation. That account withholds the status of full-fledged explanations from anything other than representations of "the relevant causal mechanism in fundamental physical terms" (130–131). This seems to be as distantly related to human understanding as are Hempel's logical deductions. Further, such a requirement apparently entails a dim view of the present state of our scientific understanding and a view at odds with what scientists take themselves to understand.

This tension between an official account of explanation and the idea that explanations generate understanding can be avoided by reversing the priorities of an account of explanation, that is, by embracing a communicative approach and emphasizing the relationship that explanations must have to explainers. The first step to this approach is simply to follow in the footsteps of all the philosophers who have acknowledged the relationship between explanations and the production of understanding. Scientific explanations are, simply put, the claims or other representations used to generate scientific understanding. Scientific understanding is to be taken in the sense developed in Chapter 4, that is, not simply as a subjective, felt sense of understanding but an epistemic achievement. Understanding in this sense is the goal of scientific explanation. Scientific explanations are designed to produce understanding in the humans who practice science and in the broader human audience for science. But this in itself is not enough to resolve the tension. The idea that scientific explanations are the means to generating scientific understanding also must be given pride of place in an account of explanation. The relationship between explanations and the cognitive needs of explainers must guide one's account of explanation. Otherwise, conflict may well arise between this idea and whatever independent requirements are placed on explanations' relationship to the world, as illustrated above with Hempel's and Strevens's accounts.

What if scientific explanations are not so tightly connected to human understanding as I have suggested? Some philosophers have challenged the idea that understanding is closely related to explanation. One concern is the idea that understanding, as a cognitive state and an epistemic achievement, is something over and beyond a successful explanation. As de Regt et al. (2009) say, "Gaining understanding through explanations is not an automatic process, but rather a *cognitive achievement* in its own right" (7). This must be right. Consider, though, which sense of explanation this claim applies to: an explanation as something out in the world, as a representation, or as an individual instance of explaining. The simple existence of explanatory facts is clearly insufficient to produce either a human cognitive state or an epistemic achievement. Both require those facts to be grasped, that is, that they bear the proper relationship to some cognitive agent. The same is true of an explanatory representation. A representation cannot generate understanding unless it is properly related to a cognitive agent, so the representational sense of explanation does not guarantee understanding. However, things are different for the communicative sense of explanation, that is, for instances of explaining. An attempted instance of explaining will not count as a successful explanation unless it generates understanding, that is, unless a cognitive achievement has occurred. I suggest, then, that successful explanation in the communicative sense is sufficient for understanding. This in turn corroborates that the communicative sense of explanation is crucial to a tight connection between explanation and understanding. Notice that none of this entails that a philosophical account of explanation is sufficient for a philosophical account of understanding, although each might contribute to the other (Grimm 2010).

A second concern with positing a tight connection between explanation and understanding arises from the idea that other sorts of things entirely, besides just explanations, can generate understanding. Lipton (2009) defends the view that understanding comes from a variety of sources other than explanations. Most persuasive of his suggestions for alternative sources of understanding is the idea that abilities or skills can directly produce tacit understanding. Lipton claims that there is no such thing as tacit explanation, even to oneself, for explanation requires explicit representation. This suggests that Lipton's focus is the representational sense of explanation, and it does seem natural to require of explanations, qua representations, that they are explicit. Explanations must be offered, even if only to oneself. Perhaps Lipton is right that, *if* understanding arises from a source other than explicit representation, there will be no explanation. But two considerations help put this gap between explanation and understanding into perspective. First, there is a wide variety of vehicles for representational content; even instances of understanding that initially appear not to arise from grasping a representation may turn out to involve representation. Second, explanations are the currency of scientific understanding. Without representation—that is, without explanation—one person's understanding cannot generate understanding in someone else. Because science is a thoroughly social endeavor, understanding without explanation is unlikely to be of much scientific significance.

The third and final concern I will consider with the connection between explanation and understanding is the idea that certain vehicles of understanding do not meet the minimum requirements for scientific explanations. We have already encountered this idea in Chapter 2, in Rohwer and Rice's (2013) contention that some highly idealized representations may be explanatory—that is, may produce understanding—but are not accurate enough to qualify as explanations. This is a version of the same concern about the necessity of explanation for understanding, but in this instance, the concern arises not out of an appreciation for the breadth of sources of understanding so much as a commitment to stringent requirements for explanation. I suspect that this kind of concern arises from expectations regarding the ontological features of explanations. Rohwer and Rice, at least, blame a representation's failure to provide an explanation on its inaccuracy. So, this is the concern that some scientific product—perhaps an idealized representation—might be illuminating, might generate understanding, but may nonetheless fall short of being an explanation in virtue of how it represents the world. Remember that understanding is supposed to include only actual epistemic achievements; the present concern is not that one might experience an errant sense of having understood but that one will truly understand without there having also been an explanation. My response is simply that any representation that can reliably generate understanding is an explanation, including heavily idealized representations. In Chapter 4, I developed an interest-relative threshold requirement of epistemic acceptability that governs when the posits of a representation are "true enough" to contribute to the epistemic achievement of understanding. Why require any more than this of our explanations?

To summarize, our scientific explanations are tailored to be the vehicles of our scientific understanding. Explanation (in the communicative sense) thus includes any systematic method of achieving understanding. Successful scientific explanations proceed via representation and connect properly to the world in virtue of their representational success. Besides communicative requirements for explanation, there are representational requirements and ontological requirements as well. Yet, as a consequence of the centrality of the communicative requirements for explanation, there is only limited sense to be made of ontic explanations existing "out there" in the world. Our explanations represent features of the world. That is the conduit whereby features of the world qualify as explanations in an ontological sense. But explanations can represent inaccurately in virtue of idealizations, and what they should represent depends in part on the audience. We should thus not place too much weight on the ontological sense of explanation. To call features of the world "explanations" without drawing attention to the representations and acts of communication from which they derive this status thus implies a mistaken source of the explanatoriness.

This view about the tight connection between explanation and understanding has two significant implications in light of the central role for scientific understanding I motivated in Chapter 4. First, it directly yokes the explanatory status of a representation to human characteristics and circumstances. As I have discussed at length, what generates understanding depends on the cognitive needs of explainers and so depends on whatever factors influence those cognitive needs. Accordingly, what qualifies as an explanation is similarly dependent. This plays a central role in the account of explanation that I develop in the second half of this chapter. Second, this view places successful explanation at the center of science, as the means for achieving science's ultimate epistemic aim. On a traditional view of science's aims, this would mean that scientists secure true explanations, and then they make use of these explanatory truths to make predictions, inform policy decisions, and so on. But this construal is, of course, not what I have in mind. On the revised conception of science's aims advanced in Chapter 4, explanations deviate from the truth when doing so aids human understanding. Explanations generate scientific understanding, but they are not concurrently true, predictively accurate, and so forth. Because of the tension among competing aims of science, the resulting scientific explanations are relatively useless for the pursuit of any other aims, including even contributing to our understanding of other causal factors involved in producing the same phenomenon.

If explanations, as the means of producing understanding, accomplish the ultimate epistemic aim of science, is this a book about scientific explanation? No; it would be misleading to say so. That scientific explanation is positioned to play these roles is due to rampant and unchecked idealization in science. Rampant and unchecked idealization is, in turn, due to the character of a science practiced by specific, limited humans to grapple with our causally complex world. The views I have advanced so far in this book do not depend on any particular ideas about the nature of scientific explanation. Instead, the account of explanation I develop in the next section follows directly from points developed in previous chapters and in this section.

5.2 An Account of Scientific Explanation

Here I outline an account of explanation that takes into consideration the communicative purposes to which explanations are put and that preserves a tight connection between explanation and understanding. There are three main elements to my account: the explanatory significance of causal patterns, the dependence of explanations on their audience, and a minimum standard of explanatory adequacy. The centrality of causal patterns to explanation derives from causal patterns' value to scientific understanding and explanations' role as the vehicles of our scientific understanding. The other two elements resolve apparent challenges created by using causal patterns to explain and by taking seriously explanation's communicative purposes. There is, of course, a vast philosophical literature on the nature of scientific explanation. I will not defend the account I develop here as the uniquely correct articulation of the nature of explanation. Instead, I employ insights from some other accounts of explanation, most notably those provided by van Fraassen (1980), Woodward (2003), and Strevens (2008b), and I will indicate where my view converges with or parts ways from these and other accounts. Nonetheless, I believe the account outlined here is a novel approach to navigating the tension between the ontological requirements and communicative requirements placed on scientific explanations, one that assimilates the insights about understanding discussed in the previous chapter.

5.2.1 THE SCOPE OF CAUSAL PATTERNS

Those who favor an ontological approach to explanation are right that any scientific explanation must have the proper connection to the world. An explanation must capture what is responsible for the explanandum, that is, the phenomenon to be explained. This means that explanations must depict dependence relations—put neutrally, factors that bear responsibility for a phenomenon's occurrence. The main issue that divides most philosophical accounts of explanation is this, the nature of the explanatory dependence relation. Are these dependence relations grounded in laws of nature, causal processes, difference-makers, causal mechanisms, or something else? Is there only one kind of explanatory dependence or multiple kinds? Is explanatory dependence ever purely mathematical? There continue to be real and significant debates about these questions, but I suspect many contemporary ideas about the nature of explanatory dependence relation(s) have gotten something right. I will thus relate what I have to say about the nature of explanatory dependence to others' ideas on the topic, emphasizing not only disagreements but also concordances, and what I think is at stake in the disagreements.

Recent decades have witnessed the ascendancy of various versions of a causal approach to explanation. This is due in part to the broad acceptance of an "explanatory asymmetry" whereby causes can explain their effects, but effects cannot explain their causes (Bromberger 1966). Famously, a flagpole's height and the position of the sun can explain the length of the flagpole's shadow, but the shadow's length and the sun's position ordinarily cannot explain the flagpole's height. Woodward's (2003) manipulability approach to causal explanation is held in high regard by many and has contributed to the dominance of one version of a causal account of explanation. Strevens's (2008b) kairetic account is another notable recent approach to causal explanation. Finally, Salmon's (1984) causal-mechanical account of explanation is part of the inspiration for the current widespread interest in mechanistic explanation (e.g., Machamer et al. 2000; Craver 2006). Despite the dominance of causal approaches to explanation, though, causation is not the only candidate account of the explanatory dependence relation. Some continue to emphasize the importance of laws or unification to explanation; others point out the possibility of mathematical or statistical dependence being explanatory.

A candidate for the explanatory dependence relation is suggested by ideas I have already motivated in this book. I have argued that the recognition of causal patterns is central to human understanding, and I have argued that explanations are the primary vehicles of our scientific understanding. This suggests that causal patterns may be the explanatory dependence relation. I develop this idea in the present section. Notice that this approach to defining the explanatory dependence relation, the ontological element of explanation, remains true to the general method outlined earlier in this chapter of prioritizing the communicative purposes of explanations, that is, what it takes to generate understanding. I posit causal patterns as the explanatory dependence relation in virtue of their importance to human understanding. As developed in Chapter 2, causal patterns are regularities embodied by some range of phenomena, where the regularities permit deviations and exceptions and the range of phenomena is limited. They qualify as causal on Woodward's (2003) manipulability approach to causation.

A causal pattern approach to explanation is a species of the dominant causal approach to explanation. The explanatory dependence relation is taken to be patterns of causal dependence, as judged by manipulability relations. This is, then, a broadly Woodwardian approach to explanation. But a causal pattern approach to explanation also does justice to at least some of the insights motivating a law-based or unification approach; in this it is similar to Strevens's account; see especially Strevens (2004). Law-based and unification approaches emphasize the explanatory benefit of generality, that is, the recurrence of the same pattern in other phenomena and across other contexts, and of expectability, that is, rendering a phenomenon un-mysterious, or to be expected. Causal patterns provide a basis for both, for their patterned nature is as significant as their causal nature. Postulating causal patterns as the explanatory dependence relation makes this dependence not simply manipulability relationships but, instead, manipulability relationships embodied, to some degree or other, across a range of phenomena. Causal pattern explanations thus identify not just features of the phenomenon to be explained but also of a general class of relevantly similar phenomena (possible or actual). In Chapter 4, I surveyed empirical research from psychology suggesting that the identification of causal relations and simple, general patterns are both significant for human understanding. Explaining with causal patterns renders phenomena expectable in virtue of providing insight into a general pattern of manipulability relations across some range of circumstances. We might call that range the *scope* of a causal pattern.

Let me say more, first, about the explanatory significance of the causal aspect of causal patterns, and, second, about the explanatory significance of the pattern aspect of causal patterns, or what I have called their scope. To begin, recall from Chapter 2 that a manipulability approach includes among causes not only the factors that changed to precipitate a phenomenon but also what I have termed "structural causes." A structural cause is a factor upon which a phenomenon counterfactually depends in the way required for the manipulability approach but that did not change in the phenomenon's actual causal history. So, causal pattern explanations include explanations that do not feature information about a phenomenon's causal history and that may not be judged to be causal from a physical causation perspective. When I first articulated this idea of structural causes, I appealed to the examples of the ideal gas law and of optimality and game-theoretic models in evolutionary biology. The latter are equilibrium models. Explanations that cite equilibrium models indicate the features of a scenario that determine equilibrium or stasis conditions and the features that allowed the system to reach equilibrium conditions, but they do not provide information about the specific processes that led to the equilibrium value in an individual instance. Equilibrium explanations thus feature structural causes upon which the phenomenon counterfactually depends, to the exclusion of actual causal history. This is a widely employed style of scientific explanation. From my example in Chapter 2: the redshank sandpiper bird prefers eating large worms, and an optimality model relating worm size to energy intake (and thus fitness) explains this preference (Goss-Custard 1977). If large worms had been historically more difficult to find (an intervention), the bird's preference, or at least the degree of preference, would be different. Yet a change in the availability of large worms need not, and likely did not, precede the evolution of the sandpipers' preference for large worms.

Equilibrium explanations are not the only causal pattern explanations that give information about structural causes. Another is explaining a phenomenon by citing a broad causal regularity of which the phenomenon in question is an instance. Here is an example adapted from Strevens (2008b). One might explain why table salt (sodium chloride) dissolves easily in water by pointing out that sodium is an alkali metal and that alkali metals have loosely bound outer electrons, which makes them prone to ionization. This explanation appeals to a causal pattern embodied by a class of entities and points out that the type of entity of interest is a member of that class. Using the ideal gas law to explain, say, the pressure of a gas in a rigid container is also an explanation in terms of structural causes. The ideal gas law posits a relationship among the amount of gas, its pressure, volume, and temperature, and that relationship structures these properties regardless of the gas's causal history. An ideal gas law explanation is the same regardless of whether a container of gas has been heated over a flame or moved to a larger container, a leaky valve decreased the amount of gas, or all or none of these things.[4]

However, causal pattern explanations need not focus on structural causes to the exclusion of causal history in the way that equilibrium explanations and causal regularity explanations do. Information about actual causal history, about the factors that changed to precipitate a phenomenon's occurrence, sometimes best captures the explanatory causal pattern. For example, consider an explanation for the high rate of mammal extinction in Australia over the past 150 years. According to Johnson et al. (2007), this is due to the decline of the dingo. As Australia's largest predator, the dingo kept the populations of smaller predators in check. As dingo populations declined, those other predator populations increased, triggering the extinction of mammal species (largely marsupials). This explanation cites features of the causal history that led to Australia's high rate of mammal extinction: the decreases and increases to population sizes that culminated in a large number of mammalian extinctions. But this is still naturally construed as a causal pattern explanation. Consider that which elements of the causal history are included in the explanation is governed by an apparent pattern the researchers are targeting—in particular, the importance of top predators to the stability of terrestrial ecosystems. Many other elements of the causal history are omitted, as they are irrelevant to that pattern (although they may well be relevant to the high rate of mammalian extinction itself, the phenomenon to be explained).

Sober (1983) and, more recently, Rice (2015) raise the concern that some equilibrium explanations do not qualify as causal explanations. Batterman (2002b) has a similar concern about some explanations that feature broad patterns. These kinds of explanation may appear to be non-causal in virtue of the fact that they do not provide information about specific causal trajectories, and adopting a physical account of causation may confirm this appearance. Recall that when I initially adopted a manipulability approach to causal patterns back in Chapter 2, I also suggested that one not inclined to go that direction may choose to think of what I call "causal patterns" as simply manipulability patterns. If one goes that route instead, she will agree with Sober, Batterman, and Rice that some causal pattern explanations are, in fact, not causal after all. These would include equilibrium explanations and at least some of the explanations I categorized above as causal regularity explanations.

Here are two concerns with this alternative approach. First, I suggested in Chapter 2 that facts about manipulability are the epistemic basis for causal ascription. Accordingly, I believe that manipulability relations are (ultimately) our basis for causal ascription, even if a manipulability analysis of the causal relation turns out to be insufficient. Second, and more to the immediate point, I wonder whether the distinction between explanations that cite what I call structural causes and explanations that cite causal history is of any significance. I have suggested that both are versions of causal pattern explanations (or manipulability pattern explanations). This indicates what I take to be the essential features of the dependence relations they cite, namely, counterfactual dependence of the sort required for manipulability and the extension of that dependence over some range of circumstances—its scope. Does anything significant about the explanatory dependence relation change when one of these causal (or manipulability) patterns involves a variable that changed as part of the causal process precipitating the phenomenon to be explained? Put another way, is the subspecies of manipulability pattern explanations that all would credit as causal different in some significant way from other manipulability pattern explanations? A prima facie reason to answer these questions in the negative is an oddity about the category I have called causal regularity explanations. Some of those cite purely structural causes, while others cite at least some causal history. This can vary even for a single explanation offered in different circumstances, to explain different phenomena. For example, we have seen that at least some ideal gas law explanations are purely structural. Others seem to capture some causal history. For instance, one might explain an increase in a gas's pressure by citing a diminished volume—say, caused by a piston—and the ideal gas law. Equilibrium explanations, causal history explanations, and causal regularity explanations: all, I suggest, are explanatory simply in virtue of citing causal patterns.

Now let me say more about the pattern aspect of causal pattern explanations. Causal patterns have two features that together constitute their explanatory significance: (1) they show how the phenomenon to be explained causally depends on one or more properties of the world, and (2) they indicate the *scope* of that dependence. Equilibrium explanations provide information about the scope of causal dependence in virtue of their setup: equilibrium models typically invoke the minimum assumptions required to generate the domain of attraction. This is one reason why evolutionary game theory models typically assume that a trait is heritable without indicating the nature of genetic influence. Employing minimum assumptions makes these models more tractable, and it facilitates their application in any circumstances when the domain of attraction obtains. It also thereby indicates the full range of circumstances in which the domain of attraction obtains, that is, gives information about the scope of the causal dependence in question. Explanations that reference a broad causal regularity (e.g., the loosely bound outer electrons of not just sodium but all alkali metals) similarly indicate the full range of circumstances in which a causal dependence obtains. And as I suggested above, causal pattern explanations that provide more specific information about causal history include only causal features that are relevant to the focal causal pattern. This enables those explanations also to indicate the proper scope of the causal pattern in question.

On this approach, then, the explanatory dependence relations—causal patterns—are not simply causal dependences but causal dependences together with the scope of dependence. Causal patterns are, after all, regularities embodied by some range of phenomena. If the explanatory dependence relation were instead taken to be simple causal dependence, this second element, scope, would be absent. To recognize the significance of this difference, consider that providing an explanation that simply described as much as possible about causal history would include wholly correct information about causal dependence, but it would fail to give correct information about the scope of dependence. If the explanatory dependence relation were simply causal, this style of explanation should be fine. In my view, it is not. On my account of the explanatory dependence relation, an explanation that included extra causal information not relevant to the explanatory causal pattern would signal that the scope of the causal dependence is more limited than it is, that is, that the pattern depends also on the extra causal factors that are cited. Causal factors not relevant to the explanatory causal pattern are not part of the explanation, even if they are important causal influences on the phenomenon to be explained. I elaborate upon and additionally motivate this point in the next section. For now, the key idea is that taking causal patterns to be the explanatory dependence relations prescribes different explanatory practices—indeed, identifies different ontic explanations—than does a simple causal view of explanatory dependence.

The explanatory significance of scope accounts for the distinctive appeal of equilibrium and causal regularity explanations. These explanations depict causal patterns, and as a result, they are at least sometimes obviously better than the alternative causal history explanations. Consider a toy example: explaining why the cup of coffee on my desk is 70 degrees Fahrenheit with the information that my office is 70 degrees and that the drink has been there for hours. This is a simple equilibrium explanation. An alternative explanation could depict some features of the actual causal history, for instance, by specifying that the eight-ounce coffee was 180 degrees Fahrenheit when placed in the 70-degree room four hours ago; how the processes of evaporation, conduction, convection, and radiation combined for a certain rate of heat transfer; and how that rate and the relative masses and heat capacities of the substances involved resulted in the beverage cooling to 70 degrees well within four hours. I submit that the equilibrium explanation for the beverage's temperature is in many instances a better explanation than the explanation that gives more information about the actual causal process. Another example, introduced above, is explaining why table salt dissolves easily in water by referencing a causal regularity among all alkali metals. As Strevens (2008b) emphasizes, explaining table salt's reactivity by referencing the loosely bound outer electrons of all alkali metals, including sodium, is better than just citing sodium's loosely bound outer electrons. In my view, this is because the former but not the latter indicates the proper scope of this pattern.

Information about the scope of a causal pattern contributes to an explanation—to understanding—in a number of ways. It indicates the circumstances in which phenomena of that type can be expected; for example, a beverage (in an open cup) will be room temperature any time it has sat undisturbed in a room for a few hours. It relates the phenomenon in question to similar phenomena that may vary in other ways; for instance, today's hot coffee and yesterday's cold tea both becoming room temperature. It is general in a way that, according to cognitive psychology, appears to be conducive to human understanding. And it contributes to our knowledge of causal patterns themselves, such as the factors upon which heat transfer depends. None of these is accomplished by causal information alone. The scope of causal dependence—the pattern aspect of causal patterns—is also crucial information.

The explanatory value of the scope of causal patterns accounts for features of explanations that various philosophers have claimed are valuable. Arguments in favor of so-called high-level explanations instead of reductive explanations appeal to the former's applicability in a range of circumstances that have certain general properties in common (e.g., Putnam 1975; Garfinkel 1981; Jackson and Pettit 1992; Sober 1999). In a similar vein, Strevens (2008b) urges that causal factors that do not truly make a difference to a phenomenon be omitted from its explanation; to this end, he argues that causal explanations should be as general as possible. Woodward's (2003) account of explanation is particularly valuable in this connection as well. For Woodward, the concept of an intervention is intended to identify causal dependencies, and the concept of invariance is intended to identify the scope of those dependencies. Finally, as I have said, acknowledging the explanatory value of the scope of dependence resonates with some insights of non-causal approaches to explanation, including the value of citing general law-like relationships, of expectability, and of unifying disparate phenomena (Friedman 1974).

Strevens's (2008b) kairetic account of explanation is also designed to accommodate the explanatory power of broad patterns, and it's worth considering how that account relates to my position here. Strevens takes the explanatory dependence relation to be causal dependence as judged by a physical approach to causation.[5] His account accommodates the explanatory value of patterns by also including a separate difference-making criterion that determines whether any given causal dependence qualifies as explanatory. All and only non-difference-making factors should be omitted from an explanation.

This creates a tension between the causal requirements for explanation and the pattern requirements for explanation, and for two reasons this in turn leads the kairetic account to reject many of what I would call causal pattern explanations. First, Strevens requires that patterns be produced by sufficiently similar microphysical causal processes to qualify as explanatory. Optimality and game theory explanations in biology fail to satisfy this requirement, as do many other equilibrium and causal regularity explanations.[6] Second, Strevens requires that explanatory patterns not omit difference-makers. For instance, he endorses equilibrium explanations only when the particular initial conditions and the subsequent trajectory toward an equilibrium are not difference-makers. But, as illustrated by Chapter 3's treatment of behavioral ecology models of cooperation, equilibrium models often use idealized assumptions to stand in for significant causal factors. These two requirements severely constrain what pattern explanations Strevens's account judges to be satisfactory (Potochnik 2011a).

The kairetic account's dismissal of some causal pattern explanations thus arises from (a) Strevens's commitment to physical causation and (b) his prohibition against the neglect of important causal influences. I have already called both of these ideas into question. Regarding (b), I argued in Chapter 4 that significant causal influences not central to researchers' immediate interests are regularly idealized to better represent a focal causal pattern and that this contributes to understanding. This renders unproblematic, and indeed unsurprising, the possibility of idealizations standing in for some important causal influences. Regarding (a), a manipulability approach to causation places no constraints on the microphysical processes that may accomplish any given causal relationship. Indeed, it even counts what I have called structural causes as full-fledged causes. If one is thoroughgoing in a commitment to a manipulability approach to causation and also recognizes that causal patterns only involve some manipulability relationships among many, this tension between the explanatory value of patterns and of causes will not arise. The explanatory value of the scope of causal dependence is ultimately why causal patterns, grounded in manipulability relations, are the explanatory dependence relations. This form of dependence reveals how the phenomenon to be explained depends on conditions that are of interest to the explainer. Causal history can be a guide to causal patterns. But many patterns, involving structural causes, are not among the changes that precipitated a phenomenon, and much causal history, even crucial causal history, is irrelevant to a given pattern. In my view, an advocate of traditional causal explanation requires an argument for why causal history per se is explanatory, independent of what this reveals about patterns of counterfactual causal dependence.

Above I discussed how a commitment to a physical account of causation leads some to question whether all manipulability patterns are truly causal. Others endorse non-causal explanations for different reasons, focusing on when explanations seem to feature wholly non-causal elements, such as mathematical explanations (e.g., Pincock 2012; Lange 2013). These kinds of arguments for non-causal explanation need not challenge a manipulability basis for causation but instead look beyond manipulability relations. I think this is a more promising basis for non-causal explanation. I have suggested that causal patterns are at the heart of science. Because of causal complexity, causal dependences abound. Furthermore, cognitive science research suggests that grasping causal relationships promotes understanding. But there is no guarantee against exceptions to this, and I have no argument that causal dependences exclusively are explanatory. If patterns in dependences other than manipulability relations are part or all of what it takes to promote understanding, those patterns are explanatory dependences. Explanations may thus sometimes represent non-causal patterns, that is, patterns not grounded in manipulability relations. Note that I expect no analogous variability around the pattern aspect of causal pattern explanations. There's (at least by and large) a causal aspect to explanations partly in virtue of the makeup of our world: because causal relationships abound. But it's purely in virtue of *our* makeup that patterns are explanatory. Any explanations generated in human science are, I submit, pattern explanations.

I want to mention one final style of argument for non-causal explanation. Bokulich (2008, 2009) has suggested that explanations feature counterfactual dependence along the lines of Woodward's manipulability relations but that these may not qualify as causal relationships. This is not due to a commitment to physical causation but because the explanations in question invoke fictional entities. Bokulich claims that such fictions can stand in counterfactual relationships but not in causal relationships. Recall her example of closed orbit theory that I discussed in Chapter 3. Classical orbits do not exist; they are fictions. So, on Bokulich's view, to invoke them to explain ionization patterns cannot amount to citing causes. Something that does not exist cannot have causal powers. However, I believe that my causal pattern approach to explanation offers a way around this challenge. Recall that causal patterns are real—embodied in phenomena, if only over some range, to some degree, and with exceptions. Invoking a fictional entity like a closed orbit or an infinite population can help capture a real causal pattern despite its fictionality. In Chapter 3, I described how this plays out for the closed orbit theory of ionization patterns. Causal patterns, not idealized representations, are the explanatory dependence relations. Idealizations, including even the introduction of fictional entities, can contribute to causal pattern explanation when they help represent real causal patterns by representing phenomena as if they had features they do not. One need not attribute causal powers to fictional entities in order to call such explanations causal.

This brings me to my final task in this section, which is to address more fully the implications of taking explanatory dependence to be causal patterns, in light of causal patterns' limitations and the essential role of idealizations in representing them. Again, as has been illustrated in earlier chapters, many causal patterns are embodied only imperfectly and have exceptions. A phenomenon may thus only approximate the pattern that, in my view, explains it. Further, some cases that appear to fall within the scope of a causal pattern explanation may constitute exceptions, that is, they may not embody the pattern at all. At least one reason for both of these situations is the fact that causal pattern explanations regularly omit significant causal influences. Representations of causal patterns are facilitated by omitting features of the phenomenon unrelated to the pattern, even causal factors that are significant in their own right. This neglect of significant factors is accomplished by incorporating idealizations. Take the example of explaining the pressure of a gas in a rigid container by appealing to the ideal gas law. This is a causal pattern explanation. The ideal gas law only ever holds approximately, breaks down entirely in some conditions, and avails itself of assumptions true of no gases. It omits mention of molecular size and intermolecular forces, implicitly idealizing them away, but both are significant causal influences. These factors result in slight deviations from the ideal gas law, and if molecules are too large or intermolecular forces too strong, a gas does not embody the causal pattern represented by the ideal gas law.

That our explanations hold only approximately and with exceptions is the cost of their pattern aspect. A causal pattern explanation that omits significant causal influences has, as a result, a greater scope. They apply—approximately and usually—to a wider range of phenomena, actual or possible. And as we have seen, broad causal patterns are the path to human understanding. The omission of significant causal influences can thus at least sometimes be justified by the increased scope of the explanation. This generates causal pattern explanations' permissive use of idealizations, including idealizations of significant causal influences. What I have already said about the relationship between phenomena and the causal patterns they embody renders this extensive use of idealization unproblematic in explanations. An idealization can be used to represent a phenomenon as if it had some property it does not, to the end of representing a causal pattern that the phenomenon (imperfectly) embodies. Thus, the ideal gas law represents a gas's molecules as points without attractive or repulsive force because these properties enable the representation of a general causal pattern.

Because explanations idealize some significant causal factors, significant causal influences on a phenomenon may nonetheless not belong in an explanation of that phenomenon. It follows that demonstrating that a factor is causally relevant does not prove its *explanatory* relevance. An explanation is actually improved by omitting causally relevant factors that are irrelevant to the explanatory causal pattern. This is different from what is generally assumed in causal accounts of scientific explanation; explanatory relevance is typically established solely on the basis of causal or difference-making relevance. In the next section, I explore the consequences of this disconnect between causal relevance and explanatory relevance. This is ultimately the reason why the communicative purposes of explanation, the centrality of which I defended earlier in this chapter, cannot be neglected. Here I have motivated causal patterns as a candidate for the explanatory dependence relation. But this isn't yet enough to indicate what, out in the world, explains.

5.2.2 THE CRUCIAL ROLE OF THE AUDIENCE

If the explanatory dependence relation were simply causal relevance, then causal relevance would guarantee explanatory relevance, and including more causal information would always improve an explanation. But for causal pattern explanation, this is not so. Properly indicating the scope of a causal pattern requires omitting information about any causal influences that do not contribute to the pattern. As I suggested at the end of the previous section, some of this omitted information may even be relevant to whether the phenomenon embodies the explanatory causal pattern. So, whether a piece of causal information belongs in an explanation depends on what causal pattern is featured in the explanation. But we saw in Chapter 2 that any phenomenon embodies multiple causal patterns, possibly countlessly many causal patterns. These include patterns in structural causes that may not involve a phenomenon's actual causal history. There are, then, many different causal patterns that could potentially explain any given phenomenon. How do we know which of these causal patterns furnishes the explanation?

One step toward an answer is obvious. Philosophers have long recognized the significance of how a phenomenon is characterized for what explains it. Put in terms of the present account, deciding on the precise explanandum, or how to characterize the event to be explained, helps determine the explanatory causal pattern. Picking up the example used above of table salt dissolving in water, one might ask why, in general, salt dissolves in water, or one might ask why, on some particular occasion, some specific sample of salt dissolved in some specific water. These are calls for explanations of a regularity and of an event, respectively. The event of some salt dissolving in a glass of water can be characterized in different ways, which gives rise to different explananda. One might ask why, on this occasion, a teaspoon of salt dissolved in the water at all, why it dissolved within some specified timeframe, or why it dissolved in exactly k seconds. These explananda are all descriptions of one event or state of affairs, but philosophers agree that they may well call for different explanations. If causal patterns explain, a causal pattern related to table salt's solubility will explain *that* it dissolved, whereas a causal pattern related to table salt's rate of dissolution will explain *when* it dissolved. In this way, specifying the explanandum can make one or another causal pattern explanatorily central.

However, specifying the explanandum doesn't fully settle the matter. Even once the phenomenon to be explained has been precisely characterized, there still can be different causal patterns of potential explanatory relevance. Another resource for distinguishing the explanatory causal pattern is also provided by existing philosophical treatments of explanation: the contrast class. The basic insight behind contrastive approaches to explanation is the idea that, as Alan Garfinkel (1981) put it, "Explanation always takes place relative to a background of alternatives" (25).[7] On this approach, an intended contrast with some counterfactual state of affairs, whether explicit or implicit, helps determine the nature of a successful explanation. For our table salt example, the idea is that we would explain why a teaspoon of salt dissolved in a glass of water (rather than settling at the bottom) with different information than we would use to explain why a teaspoon of salt (rather than a cup of salt) dissolved in a glass of water. Put in terms of my causal pattern approach to explanation, different salient alternatives—that is, different contrast classes—can also make different causal patterns explanatorily central.

Once the explanandum and contrast class are set, does this determine which causal pattern explains? I don't believe so—or, put more carefully, I believe there's no reason to think this is so. This may not be enough to distinguish explanatory causal patterns from causal patterns that are not properly situated to explain. To see why, consider the question of *why* differences in precise explanandum and contrast class influence how a phenomenon is explained. These are, it seems, expressions of the explainers' interests. The explanandum and contrast class both reflect what exactly those seeking an explanation want to understand. The explanandum reflects what features of the actual phenomenon the explanation's audience takes to be salient, and the contrast class reflects what counterfactual alternatives the explanation's audience takes to be salient. Different explananda and different contrast classes uncontroversially give rise to different explanations. But they do so,

I suggest, because they reflect the interests of those seeking an explanation, that is, what exactly they want to understand. I see no reason to think that explanandum and contrast class are the only ways in which the audience's interests influence the explanation provided.[8] If the explanandum and contrast class do not fully settle which causal pattern explains, then the audience's interests can additionally weigh in on that determination.

Here is a simple illustration of an explanation varying according to what its audience seeks to understand, even though the causal facts, explanandum, and contrast class do not change. Consider an explanation for why a teaspoon of salt dissolved in a glass of water rather than settling at the bottom. In the previous section, following Strevens (2008b), I advocated the causal pattern explanation that the sodium in the salt, like all alkali metals, has loosely bound outer electrons and is therefore highly reactive. This is sometimes, but not always, a good explanation. It is a nonexplanation for a chemist who well knows the properties of alkali salts but thought she had already added enough salt to create a saturated sodium chloride solution. Given her background knowledge, a satisfactory explanation might instead point to the fact (in this imagined scenario) that a lab tech had replaced the saturated solution with a new glass of water. These two explanations direct attention to different causal patterns embodied in the same phenomenon. One explanation appeals to the causal pattern of how alkali metals' reactivity leads to their salts' dissolvability (including this table salt). The other appeals to an obvious pattern involving a bit of the glass's causal history: the lab tech replaced the glass, and starting with a fresh glass of water is one way of guaranteeing an unsaturated solution. Both explain the same explanandum, with the same contrast class, but do so for different audiences, with different explanatory needs. Each is a wholly unsuccessful explanation for the other audience. Given not only the chemist's background knowledge but also her presumptive main interests, there is a third causal pattern embodied in this phenomenon that might provide the best explanation yet: the lab tech messed up—*again*. This is similar to an explanation that references the lab tech replacing the glass of water, but it references a broader causal pattern and may speak more directly to the chemist's only remaining interest in this explanandum.

I began this section with the question of what determines which causal pattern explains a phenomenon when the phenomenon embodies many patterns. My answer is that this is decided by what the explainers seek to understand. This is a communicative approach to explanation, in line with what I motivated in the first part of this chapter. We must first discern explanation seekers' cognitive needs before we can say how an explanation should connect to the world, that is, what causal pattern embodied in the phenomenon it should cite. An explanation's communicative purposes are, as I have said, upstream from considerations of what "ontic structures"—what causal patterns—our explanations should represent. A successful account of explanation must incorporate a formative role for particularities of the explanation seekers, a role that extends beyond settling the explanandum and the contrast class. All phenomena are influenced by a plethora of causes. All phenomena embody many causal patterns. Which of those causal patterns explain depends on the audience.

An advocate of ontic explanation may insist that, although the audience might influence what explanations are actually formulated, it does not influence the facts that explain a phenomenon, and it is these that are the ontic explanations. Here the disconnect between causal relevance and explanatory relevance is key. There is a fact of the matter about the causal influences on a phenomenon and, accordingly, whether and to what degree a phenomenon embodies a causal pattern. This is beyond the influence of those who seek to explain. But, those interests can and do influence whether an embodied causal pattern is explanatory. An explanation's audience does not shape causal relevance, but it is a key determiner of explanatory relevance. The question of whether some causal influence belongs in an explanation, and represented in what way, cannot be answered without consideration of what causal pattern explains, which depends on the explainers' cognitive needs. We thus cannot know what facts explain without considering the audience.

Because an explanation's audience influences what causal pattern explains, that audience dictates which causal influences on a phenomenon an explanation should represent and how they should be represented. The importance of which causal influences are represented is obvious. But *how* a given causal influence is represented also helps determine which causal pattern is represented. For example, one sample of table salt can be represented as some exact mass of sodium chloride, a relatively small amount of an alkali salt, a granular soluble substance, or in many other ways. The following claims represent different causal patterns that involve these different representations:

1. All alkali metals have loosely bound outer electrons and are therefore highly reactive; this is why their salts dissolve easily in water.
2. Sodium has loosely bound outer electrons and is therefore highly reactive; this is why sodium chloride dissolves easily in water.
3. A granular soluble substance will dissolve more quickly than a cube of that substance with the same mass.

Notice that the only difference between the first two patterns is their scope, the range of phenomena embodying the causal pattern, which has changed simply in virtue of how the sample of table salt is characterized. This sensitivity to how causal influences are represented means that even subtle ways in which the audience shapes how an explanation should be communicated results in a different explanatory causal pattern—a different ontic explanation.

Some research in psychology has also emphasized how audiences shape explanations. Hilton (1990) focuses on the conversational aspects of explaining. He points out that causal explanations must be not only accurate to be accepted but also relevant to the specific why-question, even when that question is left implicit. As Hilton puts it, the specific why-question helps distinguish causes from conditions. I would say instead that the why-question, that is, the explainers' focus, helps determine which of myriad causal influences are explanatory—that is, which causal pattern explains. Chin-Parker and Bradner (2010), in turn, show that background information shapes whether explainers prefer causal or functional explanations. The researchers see this as supporting a pragmatic account of generating explanations, akin to van Fraassen's (1980) account. But this result also can be viewed as background information shaping which causal pattern is explanatory, where some causal patterns hinge on what I have called structural causes and omit actual causal history. Those are the explanations Chin-Parker and Bradner identify as functional.

It may be helpful for me to say a bit more about how the present account of explanation relates to van Fraassen's (1980) pragmatic account. Like van Fraassen, I focus on the relationship between explanations and their audience, and like him, I believe the content of explanations is influenced by more than the specification of an explanandum and a contrast class. But for van Fraassen, this influence shapes the nature of the "relevance relation," what I have referred to as the explanatory dependence relation. Van Fraassen's example of different relevance relations is functional versus causal explanations, as Chin-Parker and Bradner (2010) pick up on. In contrast, I have proposed a unitary account of explanatory metaphysical dependence. All scientific explanations depict causal patterns, apparent functional explanations included. Van Fraassen and I thus differently locate the audience's influence on explanations. In my view, only one form of metaphysical dependence is explanatory, but those dependence relations, causal patterns, are so numerous for any given phenomenon that explanation seekers play a tremendous role in shaping what causal pattern explanation is called for. Framing the audience's influence as that of selecting among a few different forms of explanatory dependence doesn't do justice to the enormous variability of our explanations.

How the audience's cognitive needs influence the nature of explanations has significant implications for science. For scientific explanations, the explanation seekers' influence on what causal pattern is explanatory is largely played by what we might call the research program. A scientific research program usually involves (at least) a choice of focal phenomena, hypotheses about the types of causal factors at work or similarities with other phenomena, and a methodology, that is, a type of mathematical model, simulation, manner of investigation, or similar. Research programs are also influenced by chance factors such as what equipment happens to be available and what techniques and target systems researchers happen to be familiar with. And then, research programs can also be shaped by subtle features of the researchers themselves: their politics, their aesthetic preferences, their blindspots. All of these influence researchers' interests and, more generally, their cognitive needs. These cognitive needs bring certain types of causal relationships to the fore and ultimately determine which (real) causal patterns explain. In this way, research programs elevate what might otherwise be merely idiosyncratic features of particular explanation seekers to central influences on our scientific explanations and, thus, on the nature of our scientific understanding.

My example of salt dissolving in water is too simple to really show off how different research programs result in different explanations. Let's turn to another example, one truer to the sort of explanations actually generated in science. As I discussed above, one species of equilibrium explanation is provided by evolutionary game theory. Consider the game theory explanation for why Harris sparrows in a single flock vary in color between pale and dark (instead of being a single shade—the contrast class). Coloration is unrelated to physical strength, survival, or reproductive success, but it is related to status: dark birds almost always displace pale birds from food sources (Rohwer and Ewald 1981). Maynard Smith (1982) explains the variation in color with the hawk-dove game theory model. Because the cost of injury through conflict is greater than the value of a food source to a sparrow, this variation in coloring is an equilibrium point. Dark, aggressive birds risk injury from other dark birds but have greater access to resources; pale, submissive birds have fewer resources but avoid injury. In a mixed population at equilibrium, each is an equally successful strategy. Coloration is a badge that helps the birds divide resources while avoiding unnecessary injury.

Or, at least, that's the explanation for evolutionary ecologists interested in how frequency-dependent selection can lead to variation in a population as an evolutionary outcome. Now consider the same explanandum and contrast class—why Harris sparrows vary in color within a flock instead of being a single shade—from the perspective of a different research program. This phenomenon may also interest a developmental biologist researching phenotypic plasticity, or how the environment mediates gene expression. Such a biologist would probably want to know whether the variation in sparrow color is polyphenic, that is, whether the same genes could produce either a light bird or a dark bird in different environmental conditions. If this proved to be the case, the biologist would investigate how genetic factors and environmental variation together explain the variation in color. This explanation would feature environmental influence on gene expression in the production of bird coloration. This is a wholly different causal pattern from what is featured in the game theory explanation, but both explain the same phenomenon, characterized in the same way.[9] The two different causal patterns generate understanding in the context of different research programs.

These different causal pattern explanations represent some of the same causal influences in different ways. The game theory explanation represents what is possibly environmentally mediated gene expression with the simple assumption that feather color is heritable. The phenotypic plasticity explanation, in turn, would represent the ecological sources of frequency-dependent natural selection with the simple assumption that both feather colors persist in the population. Other biologists, with other interests, would give still other explanations, featuring different causal patterns. A population geneticist may focus on the genetic and epigenetic influences on feather color, idealizing away from both environmental mediation of gene expression and environmental sources of selection. Still another biologist might wonder if there is interplay between the pattern in environmental mediation of gene expression and the pattern in environmental sources of selection—one such biologist is Joan Roughgarden. Roughgarden's work is discussed at various other points in the text; for now, I want to merely point out the possibility of a research program focused on integrating two causal pattern explanations typically developed separately.

This raises the possibility of an alternative answer to the main question of this section: how to identify which causal pattern should feature in an explanation. Why isn't the right answer simply all of them? One might think that a complete explanation is a unified account of all the causal patterns governing a phenomenon. This is a version of Lewis's (1986) suggestion that philosophers should focus on what he calls *the* explanation—a maximally inclusive, ontic explanation. Crucially, this is the wrong answer for a causal pattern account of explanation. On this account, depicting the full gamut of causal influences in a single explanation is not only impossible but also undesirable. Introducing additional causal information not relevant to the explanatory causal pattern interferes with the explanation's scope, as discussed above. Integrating two different causal pattern explanations of a phenomenon results in a new explanation that features not both causal patterns but a third, distinct causal pattern, with a correspondingly more limited scope. To incorporate more causal details is to characterize a pattern that involves those details as well. Gone is any information about the scope of either of the two original causal patterns. Those patterns are obliterated; the integrated explanation quite literally loses the original explanatory content. This corroborates my claim from the first part of the chapter that considering only a maximally inclusive, ontic explanation is not a way around communicative requirements of explanation. A maximally inclusive causal explanation would include more causal factors than are relevant to any causal pattern that fulfills the cognitive needs of any audience. It would thus inevitably get the ontic explanation wrong.

In Roughgarden's research program mentioned just above, the explanatory causal pattern is an integrated one, involving information about the environmental mediation of gene expression along with the environmental sources of selection. The explanation featuring this integrated pattern will contain more causal information about the phenomenon being explained, but its scope is correspondingly diminished. The focal causal pattern has a smaller scope than either of the patterns in gene expression or selection alone. For Roughgarden and biologists with interests like hers, this is the better explanation. But for many evolutionary ecologists and many developmental biologists, this causal pattern is less enlightening than the game theory explanation and phenotypic plasticity explanation, respectively. It's a worse explanation for either of those audiences. Because of its diminished scope, this integrated causal pattern has diminished applicability to related phenomena. It is also expensive: Roughgarden and her collaborators have invested considerable time and effort in developing modeling techniques to represent such interactions. For researchers with a more limited focus, that would be wasted time and effort.

The persistence of distinct causal pattern explanations in science is a consequence of the explanatory significance of scope and of the continuing scientific importance of broad causal patterns. Notice that this also issues an empirical prediction about explanatory practice. My causal pattern account of explanation predicts the continuance and, indeed, the proliferation of distinct scientific explanations of any phenomenon investigated by scientists with varied research interests. I expect integrated explanations will be pursued only in special circumstances, in virtue of the expense of developing them and their focus on causal patterns with restricted scope. Those circumstances include, first, when a specific phenomenon is of particularly significant interest to scientists. An example is the evolution of large brains in hominids, a distinctive trait of our ancestors and ourselves. (Here, too, scientists' interests govern which causal pattern, albeit which more specific pattern, explains the phenomenon.) And second, when the integrated explanation features a causal pattern that recurs frequently enough to spark some scientists' interest, even with its diminished scope.

To sum up, in our causally complex world, any given phenomenon embodies a number of different causal patterns. Which of these causal patterns explains the phenomenon depends on the cognitive needs of the explanation's audience and, in particular, the research program that occasions the explanation. The research agenda, along with the interests, characteristics, and even idiosyncrasies of the scientists who develop it, crucially shapes any explanation. Uncontroversially, these features influence the precise explanandum and the contrast class, but there is no reason to expect this is the extent of the influence. I have argued that these features additionally influence which causal pattern explains a given explanandum, thereby determining what causal influences belong in an explanation and represented in what way. The influence of the research program on the content of explanations is so central that it requires a communicative approach to explanation, where the connection an explanation must have to its audience is center stage. The central role of research programs in the formulation of causal pattern explanations is, ultimately, what enables explanations to idealize significant causal influences.

5.2.3 ADEQUATE EXPLANATIONS

So far, I have suggested that the dependence relations that explain are causal patterns and that, because any given phenomenon embodies a variety of causal patterns, phenomena have many potential explanations. An explanation's audience, particularly the research program at hand, governs which of these causal patterns is in fact explanatory. Causal relevance therefore does not entail explanatory relevance. Any explanation of any phenomenon captures only a fraction of the causal determiners of that phenomenon and neglects or idealizes away the rest. These different, independent explanations for any given phenomenon is one element of the diverse aims of science and distinct deployments of individual aims discussed in Chapter 4. The aim of explaining—of generating understanding—leads to different explanations in different research programs, employing different idealizations along the way. These are different deployments of the aim of understanding.

This multiplicity of potential explanations invites the question of whether any causal pattern embodied by a phenomenon can explain that phenomenon, given the right audience. The answer is no. Consider an example. In a research program focused on genetic contributions to human behavior, one might turn one's attention to the behavior of smoking cigarettes, including the propensity to initiate smoking. And indeed, research in molecular genetics suggests that certain variants at the *BDNF* genetic locus exert a significant effect on one's propensity to smoke by affecting one's response to social stress (Amos et al. 2010). If this is right, there is a causal pattern relating some variants at the *BDNF* locus to an increased probability of smoking initiation. For the sake of our example, let's assume so. And yet it does not seem that this causal pattern succeeds as an explanation for why some individual with one of these genetic variants began the habit of smoking (rather than not forming the habit). By stipulation, in this example, the causal pattern is embodied by the phenomenon to be explained, and the research program in which the explanation is formulated attends to exactly this kind of causal pattern. But, to my lights at least, an adequate explanation has not yet been provided.

Let's formulate more carefully a description of the causal pattern in the example. The molecular genetics research suggests that certain variants at the *BDNF* locus somewhat raise the probability of smoking initiation, due to their influence on responses to social stress. Different representations of this pattern could be formulated, and the influence could be quantified, but this description is sufficient for present purposes. By assumption, the phenomenon to be explained—a particular person beginning a smoking habit, let's call her Karen—embodies this pattern. That means Karen has one of the *BDNF* variants in question and responds to social stress in a way that makes smoking initiation more likely. The problem is that there is a mismatch between what the causal pattern results in (namely, that smoking initiation is somewhat more likely), and the explanandum (namely, that Karen began a smoking habit). Whatever influence the relevant genetics have on the probability of smoking initiation, they quite obviously are not sufficient to bring about this outcome.

For a causal pattern to explain some phenomenon, it must in some sense *account for* the phenomenon. This requires, first, that the causal pattern be embodied by the phenomenon. It also requires that the causal pattern relate in the right way—whatever way that is—to the characterization of the phenomenon, that is, to the explanandum. For the molecular genetics research featured in the present example, one natural explanandum is this: heritable variation in smoking initiation among humans. A focus on why one individual, Karen, began to smoke is less fitting. Under any plausible construals of the circumstances and of genetic influence on smoking behavior, Karen's genetic predisposition to certain responses to social stress doesn't in itself account for why she began to smoke. For this reason, the causal pattern does not adequately explain. Consider that the causal pattern of certain variants at the *BDNF* locus *leading to* smoking initiation is not embodied by this or any phenomenon. In the previous section, I pointed out that characterizing the explanandum and settling on a contrast class are both influenced by explainers' cognitive needs, as is a focus on some specific causal pattern. Making incompatible choices at these junctures can result in a failed causal pattern explanation.

The additional criterion needed for an adequate explanation is, then, a match between causal pattern and explanandum or, put a little less metaphorically, for a causal pattern to account for the explanandum's occurrence. This type of requirement on explanation has been recognized by a number of different philosophers, but articulating the requirement in the right way has proven difficult. The classic deductive-nomological approach to explanation required logical entailment as the way in which an explanandum is accounted for. In a similar vein, Strevens (2008b) requires causal entailment, or the entailment of an explanandum in a way that accurately represents the causal process that led to the explanandum. But entailment is too strong. Many of the most enlightening causal patterns—the ideal gas law, predator-prey relationships, evolutionary arms races, and X and Y chromosomes' role in sex determination, to name a few—have exceptions. Alan Garfinkel (1981) proposes that explanations should feature causes that critically affect the outcome and omit overly specific details. But this approach also faces a problem: sometimes, even oftentimes, minor causal influences are center stage in the quest for scientific understanding, while significant causal factors are relegated to the role of background conditions. The smoking example used here is inspired by recent research into genetic influences on smoking behavior by Amos et al. (2010). These researchers emphasize that genetic factors are only one small part of the influence on smoking; nonetheless, their research focuses entirely on these relatively minor factors.

Requiring either entailment or a focus on significant causal influences from the explanatory causal pattern isn't the right move, but these are, I think, the right impulses. We want our explanations to focus on what's of interest to us, to respond to our cognitive needs, but we also want them to make some guarantees: that the phenomenon was bound to occur, was expectable, and, in relevantly similar circumstances, would happen again. We want our explanations to account for what they explain. In our variable and causally complex world, there is but one way to make such guarantees. Idealizations must be liberally employed. An explanation, the representation of a focal causal pattern together with idealized assumptions, must entail the explanandum, without straying too far from how the explanandum in fact came about. Call this *the requirement for explanatory adequacy*.[10] Allowing idealizations to contribute to the adequacy of an explanation significantly weakens the requirement but does not render it toothless.

The requirement of entailment is much more attenuated than it may at first appear. As I have already indicated, it is too much to expect that an explanatory causal pattern will entail the occurrence of the phenomenon to be explained. For one thing, remember that causal patterns tend to have exceptions. But it is reasonable to require that the representation of a causal pattern, along with its idealized assumptions, entails the explanandum's characterization of the phenomenon. This requirement is weaker in two ways. First, how the phenomenon is characterized can go some distance toward enabling entailment. For instance, in the example of smoking initiation, researchers may specify as the explanandum that Karen develops a smoking habit or that she was at a greater than average risk of developing a smoking habit. An explanation might entail the latter but not the former. Second, and crucially, background assumptions, including idealizations, can contribute to generating the entailment. That Karen possesses a pro-smoking variant of the *BDNF* gene does not in itself entail a greater than average risk of smoking, but that she possesses the variant and is otherwise similar to her peers does.

Karen's similarity to her peers is an uninformative background assumption, but (if accurate enough to be epistemically acceptable) it provides what is needed to entail the explanandum. This more or less stipulates that no other difference accounts for Karen's greater propensity to take up smoking. We will turn to examples that rely more significantly on idealization to generate entailment a bit further below. Notice that allowing background assumptions to contribute to an explanation's adequacy makes it so that the context for the explanation, including the audience's background knowledge and expectations, can help secure explanatory adequacy by providing shared background assumptions that may go unmentioned. If the researchers already know that Karen was chosen for study in virtue of her averageness, the discovery of her *BDNF* variant may well explain her greater propensity to take up smoking.

I also said that, for explanatory adequacy, a causal pattern explanation must not stray too far from how the explanandum in fact came about. I don't intend this as an additional requirement; it is guaranteed by elements of my view already in place. Recalling these elements here helps to motivate my proposed requirement for explanatory adequacy. There are two ingredients, corresponding to the represented causal pattern and the requisite background assumptions. First, as I have already discussed, any explanatory causal pattern must be embodied by the phenomenon it explains.[11] And second, the idealizations and other background assumptions that enable the representation of this causal pattern must successfully represent as-if. The embodiment of causal patterns and representation as-if were both discussed in depth in Chapter 2. The latter requires idealizations to be functionally similar to the elements of the phenomenon they represent in a limited way needed for the focal causal pattern. An example is the humble assumption, discussed just above, that

Karen is otherwise similar to her peers. Representation as-if enables significant causal factors that are not central to the current research program to be set aside but still taken into account. In this way, idealizations are used to bring a causal pattern explanation into alignment with the totality of causal facts, so that it counts as an adequate explanation.

Both ingredients, the embodiment of the explanatory causal pattern and idealizations that successfully represent as-if, are assured via the satisfaction of the variable requirement of epistemic acceptability outlined in Chapter 4. Recall that this requires a posit to not diverge significantly from the truth, taking into account the posit's role in a representation and the epistemic purpose of the representation. I have suggested that any posit contributing to understanding must be epistemically acceptable in this way. In light of the tight connection between explanation and understanding, it follows that any posit contributing to an explanation must satisfy this requirement of epistemic acceptability. And so, posits central to representing a focal causal pattern in some phenomenon must accurately represent the causal factors contributing to this pattern. This ensures the phenomenon embodies the focal causal pattern. Idealizations, in contrast, must simply not go too far wrong. To qualify as epistemically acceptable, they simply must help the representation approximate the behavior of the phenomenon in relevant respects. In other words, idealizations must be functionally similar in certain respects to what they represent as-if. So, when all posits of an explanation satisfy the variable requirement of epistemic acceptability, this ensures that the phenomenon embodies the explanatory causal pattern and that none of its neglected features interferes dramatically with that pattern.

Here is an illustration of how idealizations help secure explanatory adequacy. Recall from above the hawk-dove game theory explanation of why Harris sparrows vary in color between pale and dark. Dark birds are aggressive and risk injury but have greater resource access; pale birds are submissive and avoid injury but have fewer resources. This variation in coloring is an equilibrium point because each is an equally successful strategy. This game theory model does not by itself entail that Harris sparrows vary in color—for that, a long list of assumptions, many of them idealizations, is also required. One of those assumptions is that the traits of paleness and darkness propagate to the degree of their relative success. This is an idealization; it is technically false of all genetically influenced traits. Instead, bird color ratio in one generation and in the next share a common cause, namely, genes that both influence feather color and propagate themselves. This idealization is a way of taking into account—without accurately representing—the role of genetic influence on color and on genetic transmission. The same is true for the idealization of an infinite population size, as well as the many other assumptions required for evolutionary game theory explanations. All of these assumptions are needed to generate the domain of attraction—that is, the range of conditions that would lead to the equilibrium ratio of pale to dark birds—and to locate Harris sparrows inside that range. When coupled with all of these idealizations and other background assumptions, the hawk-dove game does entail that Harris sparrows vary in color between pale and dark. This is an adequate explanation.

Let me mention a couple other elements of my treatment of this example. First, meeting the requirement for an adequate explanation does not guarantee a *good* explanation. The aim of my criterion for explanatory adequacy is to establish a benchmark that must be met by any causal pattern explanation. It is not by itself a sufficient condition for successful explanation, but only a necessary condition to be met by any otherwise satisfactory explanation. In light of the other elements of the account of explanation developed in this chapter, the hawk-dove explanation will only succeed if the causal pattern it characterizes is embodied by this evolutionary episode and if the biologists seeking explanation seek an evolutionary ecology explanation. The former amounts to the explanation qualifying as epistemically acceptable, and the latter is determined by attending to the research program at hand. Second, notice that other factors, such as genetic drift, may have shifted the ratio of pale birds to dark birds away from what the game theory model predicts. If so, this does not undermine the explanation described here, but it does mean that the hawk-dove game does not adequately explain (i.e., does not entail) the exact ratio observed but only a more general characterization of the outcome. If the predicted ratio does obtain (at whatever level of specificity the explanandum characterizes it), then on my account the hawk-dove game can explain this ratio, even though it idealizes away drift and other significant causal influences.

So, my proposed criterion of explanatory adequacy requires that an explanation—the representation of a causal pattern together with its idealizations and other background assumptions—entails the explanandum. But why think such a requirement should be endorsed, other than the intuition I tried to motivate regarding some explanations coming up short? In my view, this form of explanatory adequacy is needed for an explanation to contain the proper information about scope. For the example hawk-dove game explanation, the idealizing assumptions that help generate the entailment also help generate the domain of attraction. They thus help indicate the conditions in which the equilibrium value results, in other words, the scope of this causal pattern. Requiring that a causal pattern explanation, including its idealizations, entails the explanandum thus indicates the limits of the scope of a causal pattern. This also demonstrates that the explanandum is within that scope, thereby showing that the phenomenon, as characterized in the explanandum, was to be expected. Even low-probability events and events that are explained with only enough causal detail to confer on them a low probability can be made expectable—that is, can be shown to be within the scope of a causal pattern—by appropriately tailoring the explanandum and employing the right idealizations. (This presumes the phenomenon embodies the pattern in question.) That an individual won the lottery is never explained by the fact that the she bought a single ticket. But, in some cases, the explanation of a person winning the lottery is that she bought a single ticket and got extraordinarily lucky. The invocation of luck is needed here to indicate the extreme limitations of the causal pattern where buying a lottery ticket leads to winning. In this case, the limitation is not in the scope of the pattern so much as in the plenitude of exceptions.[12]

The failed explanation of Karen's smoking initiation is straightforwardly ruled out by this criterion for explanatory adequacy. Possessing a variant at the *BDNF* genetic locus does not, by itself, entail an individual's smoking initiation. The causal role of that variant is highly sensitive to circumstances like social norms and tobacco availability, circumstances that the explanation in no way takes into account. This example is an obviously bad explanation (of the given explanandum), but it can be tricky to establish whether an attempted explanation satisfies the requirement of explanatory adequacy. This is especially so because the satisfaction of that requirement rests in large part on causal influences that are not central to the research focus, causal influences that are neglected to the extent that the researchers' approach allows. The need for additional background assumptions and idealizations can, and regularly does, come to light after an explanation has been accepted as adequate. Similarly, explananda may be subtly adjusted to accommodate previously unappreciated causal influences. An example of both is the recognition of genetic drift as an important causal influence on evolutionary outcomes. In his classic overview of evolutionary game theory, John Maynard Smith (1982) discusses many of the assumptions required for that approach, but he does not mention the need to assume an infinite population. As the influence of genetic drift has gained appreciation, the need for this assumption has been made explicit. This is also one reason why explananda for evolutionary game theory models tend to allow for some deviation in trait value due to influences other than natural selection.

The three ideas I have motivated in the second half of this chapter together constitute an account of scientific explanation that pays heed to the relationship that explanations must have to human explainers. This account grants the status of full-fledged explanations to those representations that can give rise to human understanding as specified in Chapter 4. My account of scientific explanation can be summarized as follows. The explanation of some explanandum E must (1) represent a causal pattern embodied in the phenomenon characterized by E, including a causal dependence and the scope of that dependence. Further, (2) this causal pattern must be of central interest to the research program in which the explanation is generated, and (3) the representation of the pattern and its background assumptions must together entail E. Both (2) and (3) are partly accomplished by the specification of the phenomenon to investigate, its characterization as an explanandum, and perhaps a contrast class. But both also require more than this: (2) of the relationship an explanation must bear to its audience and (3) of the relationship an explanation must bear to the world.

6

Levels and Fields of Science

In the preceding chapters, I have developed an account of the aims and practice of science that features, at its core, the idea that idealization is rampant and unchecked. Here I discuss implications of these ideas for philosophical views about levels of organization and field divisions in science. I conclude that the phenomena under investigation in science are not organized into discrete levels. Relationships in composition, scale, metaphysical determination, and causal dependence have variously been linked to levels. But idealized representations in the service of diverse aims should lead us to question the idea that different targets of investigation are related in metaphysically significant ways. In turn, causal complexity should inspire caution about limitations placed on relationships of causal influence, like the expectation that they are limited by compositional relationships. These ideas also motivate a view of scientific fields and subfields as reflecting different simplified strategies to pursuing understanding, prediction, and control of causally complex phenomena. Field divisions thus should not be taken to reflect facts about the metaphysical status of the entities under investigation, be it their levels of organization, realization relationships, or relations of causal influence.

In section 6.1, I survey how the concept of levels of organization is customarily deployed in philosophy and how it has been invoked in philosophical debates about reduction, levels of scientific explanation, emergence, realization and multiple realization, and others. I then more briefly consider the role this idea plays in a few areas of science. Then, in section 6.2, I argue that there is no general conception of levels of organization that is of scientific significance. I also explore how the relationships traditionally conceived of in terms of levels can more fruitfully, and less confusingly, be interpreted. This involves distinguishing among a range of commonsense notions—including composition, spatial scale, and direction of causal influence—that, when conflation is avoided, can more effectively do the work of the old, problematic concept of levels.

Finally, in section 6.3, I address the status of science's fields and subfields in the absence of levels. Although fields of science have classically been taken to reflect discrete levels of organization, they are more naturally understood as a practical division of labor with little metaphysical significance. Field divisions simply reflect the fact that the phenomena of human scientific concern are complex and best addressed with highly simplified approaches, employing characteristic idealizations. This results in many different approaches that resist integration. This view nicely accounts for the array of fields and subfields observed in current science, as well as how those change over time. It also entails a different view of the relationship among fields of science, classically thought of as the question of the unity or disunity of science. No longer are metaphysically significant relationships, such as realization, anticipated among the entities addressed by different fields. Nonetheless, there is a kind of unity among fields. The widespread hunt for causal patterns and the continuing significance of idealized representation limit the ability of individual scientific investigations to provide the evidence needed to corroborate their own theories and explanations. This results in epistemic dependence among different fields and subfields. Accordingly, both division of labor and cross-connections among those separate projects are essential to science.

6.1 Levels in Philosophy and Science

The concept of hierarchical levels of organization is commonplace in science and philosophical treatments of science.[1] In the sciences, hierarchical notions are applied to a few different types of relationships, but there seems to be one primary use of the concept in philosophy. This is a *compositional* hierarchy, based on the idea that material composition gives rise to discrete levels. Subatomic particles uniformly compose atoms, which compose molecules; cells compose tissues, which compose organs, which compose organisms; interbreeding organisms compose populations, which compose communities, which compose ecosystems; and so on. Because of its prominence in philosophy, the notion of a compositional hierarchy is my main focus in this chapter, although I discuss connections with other conceptions of hierarchy where they are relevant.

The basic idea of a compositional hierarchy is that higher-level entities are composed of (and only of) lower-level entities and that this gives rise to discrete levels of entities, fully ordered both by size and by compositional relationships. These are the classic "levels of organization." This concept of levels is typically deployed in a way that incorporates stronger claims than simply the ubiquity of part-whole composition; it is often also taken to involve stratification into discrete and universal levels of organization. Not only are compositional relationships ubiquitous. It is also the case that we see the same types of parts composing the same types of wholes in a wide range of systems. Atoms, for example, come in limited variety, and all of these bear various things in common with one another. Most of these atoms combine, following predictable rules, into molecules of a broader variety. Some molecules— the organic ones—can combine into cells, which are more various still, yet nonetheless can be classified into types, bear certain things in common with one another, and can combine to form more complex entities at the next higher level of organization, organic tissues. These levels are discrete insofar as entities at one level uniformly compose entities at another level (if they compose anything at all), and they are universal insofar as these same types of composition recur in all different circumstances (provided the circumstances enable any composition at all). It is also often assumed that levels are nested, that is, that an entity at any level is composed of aggregated entities at the next lower level. Intuitively, a population is simply an assemblage of organisms that bear a particular relationship to one another, just as a molecule is simply an assemblage of atoms that bear a different particular relationship to one another. It seems to follow that entities at higher levels are always larger and more complex than entities at lower levels and that the compositional relationship is transitive.

Something like this conception of levels of organization makes an appearance in a wide range of philosophical debates, including those surrounding the mind-body relationship, the unity or disunity of science, explanatory reductionism or antireductionism, ontological reductionism or antireductionism, emergence, and others still. Many different implications of this basic idea of levels have been posited in the context of those diverse philosophical discussions. And yet, the roles the concept of compositional levels plays in these various debates are seldom developed carefully and are even more rarely articulated together, as part of a unified discussion of the nature and implications of levels of organization. For this reason, I devote some space here to articulating the range of implications that levels have variously been taken to have.

James Feibleman (1954) outlines an expansive view of the significance of levels of organization that prefigures many later, more specific treatments. He claims that entities at each successive higher level of organization possess new properties not belonging to their components, for they are particular organizations of their lower-level parts. And yet, on the other hand, any high-level object "depends for its continuance" on its lower-level elements (66). A living cell has the property of self-replication, whereas the molecules comprising it do not, but there would be no cell without molecules. Feibleman also claims that complexity always increases as levels are ascended, while rate of change always decreases. This is related to his idea that organization at a given level is accomplished by a mechanism at the level below, for the purpose it fills at the level above. Feibleman's assertions about the significance of the compositional hierarchy are especially blunt, but they make explicit several attributes that many philosophers—both then and now—expect of levels.

Consider first Feibleman's comments about the relationship between higher-level and lower-level properties, namely, the existence of distinctive higher-level properties and their dependence on the lower level. The idea that higher-level entities possess novel properties is the basis of the idea of emergence. Emergent properties are generally taken to be high-level properties that cannot be predicted, explained, or reduced by lower-level properties, insofar as the organization of components is crucial to the properties' emergence (Kim 1999; Mitchell 2012a). The existence and causal relevance of emergent properties is disputed, as is the proper articulation of emergence. Less controversial is the idea that higher-level properties depend on lower-level properties; even proponents of emergentism generally grant that all properties of any (physical) system are determined by the properties of its microphysical components. This is usually cashed out in terms of realization or supervenience (Horgan 1982; Kim 1999). Notice that this kind of metaphysical determination is distinct from causal determination. The idea is that there can be no change in a higher-level property without a corresponding change in one or more lower-level properties. This seems to follow from the compositional relationship of parts to a whole, so long as the parts are all there is to the whole (i.e., barring the possibility of special introductions at higher levels, such as an élan vital or nonphysical soul).

Another of Feibleman's claims that resonates in a useful way with more recent discussions of levels in philosophy is the idea that change in an entity at a given level is accomplished by a mechanism at the level below and that the purpose of that change is established by the next level above. Some philosophers take mechanisms to be crucial to making sense of the causal significance of levels. Machamer et al. (2000) characterize mechanisms as entities or activities that produce regular changes, and they claim that mechanisms form nested hierarchies. That is, lower-level mechanisms have dedicated roles as parts of higher-level mechanisms. Craver and Bechtel (2007) in turn argue that mechanisms are key to making sense of all interlevel causal claims. They suggest that causal relationships only occur within a single level, so claims of a causal relationship between entities at different levels actually should be understood as partly a causal claim and partly a constitutive claim. These and other advocates of mechanistic levels articulate a specific basis for hierarchical organization that's distinct from simple composition. However, some also suggest that their favored characterization is in fact a version of the traditional compositional hierarchy. According to Craver and Bechtel (2007), "Levels of mechanisms are a species of compositional, or part-whole, relations" (550). They cite as evidence for this some of the features of levels identified here: that levels of mechanisms are ordered by size, with lower-level entities smaller than higher-level entities, and that levels of mechanisms have slower actions and greater complexity at higher levels.

A further type of significance that has been attributed to levels of organization relates to scientific explanation. According to explanatory reductionists, higher-level events and regularities should be explained by demonstrating how they arise from lower-level events and regularities (classically, Oppenheim and Putnam 1958; Nagel 1961; Hempel 1966). This thesis is justified with an appeal to the metaphysical dependence of higher-level entities on lower-level entities, namely, realization or supervenience. If all phenomena are fully determined by entities, causal relationships, and laws at lower levels, then perhaps uncovering lower-level truths—and their relationship to higher levels—is key to explaining higher-level phenomena. Arguments against explanatory reductionism also often rely upon hierarchical organization but to opposite effect. Those appeal to emergent properties at higher levels, that is, properties that do not reduce to lower-level properties (Mitchell 2012a), or to multiply realizable properties, that is, properties that supervene on any one of many distinct lower-level properties (Fodor 1974; Putnam 1975; Garfinkel 1981). More generally, Wimsatt (2007) claims that there is a "level-centered orientation of explanations" (214).

So far I have focused on views about the significance of levels of organization in entities. But these ideas about the hierarchical organization of entities have also been yoked to ideas about how the scientific investigations of these hierarchically organized entities are stratified into levels. Indeed, hierarchical organization as a metaphysical thesis is often not clearly distinguished from hierarchical organization as a thesis about the fields of scientific investigation. Discussions of levels of explanation in particular tend to combine assertions about hierarchical organization of entities and of scientific knowledge. If entities are organized in a compositional hierarchy, then, it may seem, so too are the fields that investigate those entities. This gives rise to the "pyramid" view of the sciences that is sometimes invoked in philosophy; see especially

Oppenheim and Putnam (1958). On this view, microphysics, the investigation of subatomic particles, is the lowest level of science; chemistry somewhat higher; cellular biology higher still; and macroecology and social sciences near the top of this hierarchy of fields. Microphysics is supposed to be the most general, as everything is composed of subatomic particles, whereas the social sciences, for example, only regard a small subset of ordered subatomic particles, namely, human societies. Views opposed to this picture are usually taken to assert the disunity of science, understood as the absence of this kind of an ordering of scientific investigations (Cartwright 1999).

Finally, some explicitly or implicitly consider lower-level scientific theories to be epistemically more secure than higher-level scientific theories. This is a claim about the hierarchical organization of science rather than of entities. An example of this view is Oppenheim and Putnam's (1958) suggestion that all scientific investigations ultimately may be vindicated by demonstrating their foundation in microphysical law. Such notions of the epistemic value of the lower level are seldom explicitly distinguished from claims of the explanatory value of the lower level. Yet even avowed antireductionists about explanations sometimes assume that physical theory is somehow epistemically privileged over theories in the biological and social sciences. It seems to be a common assumption that if such theories conflict, the higher-level theories should be rejected in favor of the lower-level theories. Claims to this effect are not often defended in writing and occur more often in casual conversation. However, Ladyman and Ross (2007) explicitly state this in the form of a general principle:

> Special scientific hypotheses that conflict with fundamental physics, or such consensus as there is in fundamental physics, should be rejected for that reason alone. Fundamental physical hypotheses are not symmetrically hostage to the conclusions of the special sciences. (44)

This principle asserts a strong version of the epistemic priority of lower-level scientific investigations.

A similar conception of compositional hierarchy is prominent in many areas of science as well. Consider the field of ecology. Nearly all ecology textbooks use the idea that biological organization is hierarchical as an organizing principle. Two prominent textbooks (Ricklefs 2008; Molles 2002) both have an introductory chapter with a description and visual representation of the levels of organization, then order their book sections according to those levels. Figure 6.1 is a reproduction of one of the first figures appearing in yet another textbook (Sadava et al. 2009). Editions of these textbooks have been around

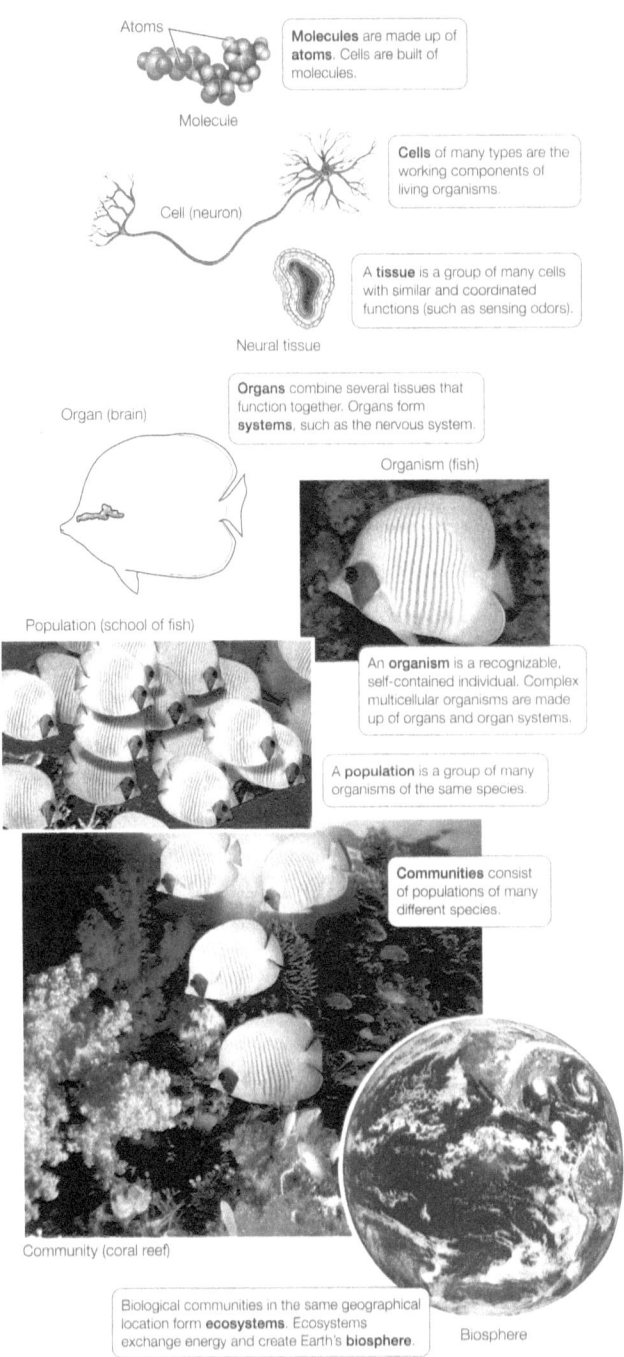

FIGURE 6.1. One of the first figures in most ecology textbooks depicts hierarchical levels of organization and asserts the relevance of levels for the study of biology, especially ecology. This is Figure 1.6 from Sadava et al.'s *Life: The Science of Biology* (2008).

for decades and have been used to train most practicing ecologists today. The same also appears to be true of cell biology textbooks. Nor is the concept of hierarchical levels of organization limited to promoting basic understanding in students and non-experts. According to Lidicker (2008), "The nested levels of organization hierarchy has been widely adopted by biologists" (72). Commitments to hierarchical organization are also common in the social sciences (Epstein 2012) and in classical physics (Rueger and McGivern 2010).

There are some striking parallels and overlaps in how the concept of levels of organization is treated in philosophy and in science. For example, in virtue of the assumption that levels are nested, the biological hierarchy is sometimes referred to as "the pyramid of life" (Lidicker 2008). This is at or near the apex of the pyramid that has been used to represent the classic reductionist conception of the whole of science. Epstein (2012) discusses how a pyramid view is implicit in conceptions of hierarchical composition in the social sciences, for example, in the idea that companies are composed of divisions, which are in turn composed of employees. Similar to the philosophical position of explanatory reductionism, many scientists look for explanations at lower levels of organization. In ecology, this takes the form of seeking explanations from population dynamics (MacArthur 1968; May 1976; Wilson and Bossert 1971), from physiological and behavioral processes (Schoener 1986), or even from chemical and physical principles (West and Brown 2005). In the social sciences, a commitment to "ontological individualism" is common; this is the idea that social properties are entirely determined by the properties of individuals.

The ideas that higher levels of organization are marked by greater complexity and slower rates of change, discussed above in connection with Feibleman (1954), are both found in ecology as well. Jagers op Akkerhuis (2008) offers the following brief characterization of the idea that complexity increases with hierarchical organization:

> The organization of nature is profoundly hierarchical, because from its beginning, interactions between simple elements have continuously created more complex systems, that themselves served as the basis for still more complex systems. (2)

This couples the idea of nested composition with causal production, such that higher-level entities are automatically taken to be more causally complex than their lower-level parts.[2] O'Neill et al. (1986), in a very influential book on the field of ecology, rely on similar ideas to justify their notion that hierarchies can be distinguished by their different rates of change. They claim that higher levels will change more slowly than lower levels. Their reasoning seems to be that lower-level interactions amass to result in higher-level behaviors, so the latter must be slower than the former. Although they acknowledge difficulties with the classic view of compositional levels of organization, O'Neill et al. do not clearly distinguish their conception of hierarchical organization from the classic view, and their main examples are also commonly taken to be examples of nested compositional hierarchies. These authors further claim that "any attempt to relate a macroscopic property to the detailed behaviors of components several layers lower in the hierarchy is bound to fail due to the successive filtering" (80–81). This seems akin to philosophical antireductionism about properties.

Biological levels of organization also relate to some central ideas in evolutionary theory. The "major transitions" in evolutionary history, as introduced by Maynard Smith and Szathmáry (1995), are basically the emergence of new, higher levels of biological composition. At certain points in the evolutionary past, independently replicating entities joined forces in chromosomes, prokaryotic cells merged to form the first eukaryotic cells, cells joined to form the first multicellular organisms, and organisms organized themselves into colonies or social groups. Maynard Smith and Szathmáry focus on general causal implications of these transitions: how larger entities tend to take over causal processes, like replication, formerly accomplished by the smaller entities, and how smaller entities nonetheless retain some autonomous causal influence, for instance, by possibly disrupting developmental processes. The issue of levels of selection is also relevant here. In the present context, that can be construed as the question of where in the biological hierarchy natural selection exerts its causal influence: on genes, individuals, or both and groups of individuals as well. These connections with evolutionary theory suggest that the evolutionary process is causally responsible for some of the compositional relationships we observe. This, in turn, may give us reason to think these levels are general (at least in some phenomena) and that they have causal significance.

This discussion of the implications attributed to the compositional hierarchy is not exhaustive, but it does capture many of the prominent ideas formulated in philosophy and in science about the significance of levels. These include claims about the metaphysical significance of hierarchical levels, from the very idea of discrete, stratified levels of entities to relationships that have been posited among the entities and properties at these levels, including emergence, realization, and supervenience. There are also claims about the causal significance of levels. These include ideas put forward about the relative complexity and rates of change of different levels, as well as the conception of hierarchy in terms of levels of mechanisms. Then there are claims about the explanatory and evidential significance of levels and the unity or disunity of science. For the most part, these pertain to theories or investigations at different levels rather than to entities at different levels of organization. For all of these claims about the significance of hierarchical organization, some version of a universal stratification into compositional levels is assumed.[3] Yet this assumption often remains in the background; as Kim (2002) points out, the conception of levels employed is seldom made explicit.

6.2 Going without Levels

As we have seen, the classic formulation of levels of organization in philosophy of science is as a compositional hierarchy among entities, and something like this formulation is common in parts of science as well. I have identified several implications this conception of levels of organization has sometimes been taken to have. In this section, my first task is to develop criticisms of all but the most basic conception of the compositional hierarchy. Even that basic conception is more limited than commonly assumed. I then outline a variety of ways in which thinking in terms of levels does make sense. This will demonstrate that my negative conclusions about levels of organization and their supposed implications are not overly radical. There is no need to lose the baby along with the bathwater. On the other hand, I also argue against adopting any general concept of levels of organization, even suitably revised. It is better instead to recognize a range of partially overlapping relationships that entities may bear to one another.

6.2.1 AGAINST HIERARCHY

Let's first consider the very idea of levels of organization. Claims about hierarchical organization often appeal to the ubiquity of part-whole composition. Indeed, the existence of stratified levels depends upon not only the ubiquity but also the uniformity of part-whole composition. As Markus Eronen (2013) points out, composition is merely a relationship between parts and wholes. It does not determine the relationship those parts have to other parts or the wholes to other wholes. The assertion of stratified levels of organization thus requires an additional assumption beyond simple part-whole composition; something like the idea that similar entities are similarly composed. For strata to emerge, atoms must always compose molecules (if they compose anything at all), populations must always compose communities, and so on. This is the uniformity required for compositional relationships to give rise to levels.

But the uniformity of composition needed for stratified levels simply does not exist. In our causally complex world, where causal contributions are numerous and vary tremendously, causal participation in a system or entity comes in many forms. Guttman (1976), a biologist, forcefully made something like this point.[4] He provides a variety of examples demonstrating that objects at some level n are in general not composed exclusively of objects at level $n-1$. For instance, ecosystems are said to be composed of populations or communities, but individual molecules, such as molecules of food waste, are also an important component, with distinctive causal contributions. Food waste molecules are essential to the movement of nutrients through an ecosystem and the retention of water outside of organisms. Similarly, tissues are only partly composed of cells; also crucial are the macromolecules that hold the cells together. This pattern persists at lower "levels" as well: polymer molecules are composed of monomers, but individual ions are also an important ingredient that play an important role (Guttman 1976). Thus, while every whole may be composed of smaller parts, it is certainly not the case that every whole is composed of only parts at the next lower level (as traditionally conceived). Nor is it the case that each *type* of whole is composed of all and only the same types of parts. Consider organisms. Big, lumbering organisms like us are composed of organ systems, as well as smaller components, such as cells that act individually, like blood cells' role in oxygen transport. At the other extreme, single-celled organisms are not composed of cells at all; they are composed of cell parts, such as organelles. There is also an array of different compositional configurations between these extremes, as well as some entities whose status as organisms is a matter of dispute (Wilson 2008). These two failures of uniformity call into question the idea that there are discrete levels of organization.

Similar limitations, also arising from causal complexity, apply to the idea that higher-level properties are realized by lower-level properties. It may be that every higher-level property is realized by one or more lower-level properties, but this does not define or correspond to stratified levels of organization. First of all, each type of higher-level property need not be realized by the same types of lower-level properties. This follows from the well-appreciated phenomenon of multiple realization. Many higher-level properties are realized by any of a variety of lower-level properties, and these properties may occupy different (presumed) levels of organization (Melinda Fagan, in conversation). For instance, an organism's property of camouflage is realized by an array of lower-level properties. For many animals, camouflage is accomplished by pigmented cells that disguise through coloration. Other instances involve morphological structures of the animal or materials in the environment. The properties of cells, of larger parts of animals, and of nearby materials variously realize organism camouflage by matching backgrounds, concealing shadows, obliterating forms, disguising motion, enabling masquerade as other objects, or creating other perceptual effects (Stevens and Merilaita 2009). Such multiple realization means that any number of types of properties, of any number of types of objects, at any number of so-called levels may realize a given higher-level property.[5]

There is also no guarantee that higher-level properties are realized by a well-defined set of lower-level properties (Epstein 2012; Mitchell 2012a). Even an individual occurrence of some property may be realized by a complicated combination of other properties. I have termed this "complex realization" (Potochnik 2010b). The properties that jointly realize a given instance of a higher-level property may have little in common with one another, other than all contributing to the occurrence of that property. Accordingly, these contributors may never be grouped together as an object of scientific study. To return to the example of camouflage, consider what properties determine an instance of, say, a fish having the property of being camouflaged as a fallen leaf on a stream bed. Relevant properties of the parts of the fish include the properties that determine its overall shape, that it is positioned on its side, and the colors and arrangement of its pigmented cells. Properties of the environment are also relevant: the presence of leaf litter in the stream, as well as the leaf litter's heterogeneity, size, and color. Finally, properties of the predators also matter for this to achieve camouflage: predators must be diurnal, aquatic, and hunt by sight (Sazima et al. 2006). The realizer of this case of camouflage—that is, the set of properties that metaphysically determines the presence of the camouflage property in this instance—is a complicated set of properties belonging to fish parts, elements of the fish's environment, and predator populations. This set of properties is not naturally ascribed to entities at a single level of organization, nor are they targeted by the same scientific investigations.

This means that realization relationships cannot be used to distinguish among levels of organization. Multiple realization and complex realization give rise to groupings just as complex and unstratified as those based on part-whole composition. It is also worth noting that groupings based on realization and those based on composition are largely distinct. Furthermore, realization relationships have little significance for the relationships among different scientific pursuits. Scientists engaged in different types of research often focus on different types of entities and, as we have seen, different causal patterns. Some of those entities bear a part-whole relationship to one another. But, whenever there is complex realization, which is often, the properties and relationships these research projects focus upon are not related by realization. That is, the realizers of some property of interest are hardly ever grouped together as a direct target of scientific investigation. This undercuts the scientific relevance of realization relationships. I say more about the relationships among fields of science later in this chapter.

To summarize, composition may be universal, in the sense that every extant whole is composed of proper parts, and realization may be universal, in the sense that every nonfundamental property is accomplished by some other combination of properties. But neither gives rise to discrete, stratified levels of organization. Levels are not discrete in that it is not the case that an object taken to be at some level n is composed of only parts at level $n-1$. Levels are not stratified in that it is not the case that all objects taken to be at level n are composed of parts from the same levels $j \ldots k$. Similar things can be said of realization relationships, which tend to be multiple and complex. Because of multiple realization, the presence of some nonfundamental property provides little or no information about its realizers, even the level(s) at which its determiners are found—the realizers are highly variable from instance to instance. Because of complex realization, the nature of the realizer in a given instance is even obscure. There may be a long list of relevant properties, and these likely do not belong to the same kinds of entities or even to parts of the object with the focal property. Even if realization is universal, it is not associated with the hierarchical stratification of objects or their properties.

Now consider the supposed explanatory significance of hierarchical stratification. The idea of supervenience or realization is at the heart of explanatory reductionism, that is, the view that higher-level events and regularities should be explained by showing how they are metaphysically determined by lower-level events and regularities (classically, Oppenheim and Putnam 1958; Nagel 1961; Hempel 1966). In a common countermove, other philosophers have used multiple realization as a basis for arguing against the explanatory priority of the lower level (e.g., Fodor 1974; Putnam 1975; Garfinkel 1981). For multiply realized properties, higher-level regularities exist despite lower-level variation. Antireductionists use this as grounds to dispute the idea that understanding of higher-level phenomena is provided by lower-level information. But complex realization—higher-level properties' realization by an obscure combination of lower-level properties, as introduced above—undermines both of these conclusions about the significance of levels for scientific explanations. The properties featured in explanations judged to be lower level are seldom if ever the realizers of the properties featured in explanations judged to be higher level. Consistent with the account of explanation I developed in Chapter 5, so-called lower- and higher-level explanations simply capture different causal patterns. Explanatory reductionists and antireductionists alike thus err in using realization as a way to evaluate the relative value of explanations formulated at different levels. Realization and multiple realization have little relevance for the comparison of explanatory strategies. Explanations may be categorized as lower or higher level, insofar as they refer to smaller or larger objects—perhaps even to parts or wholes. But this distinction has little importance since it cannot provide a significant basis for comparison. (See Potochnik 2010b for a more extended treatment of this idea.)

The original goal of explanatory reductionism was epistemic in nature. Oppenheim and Putnam (1958) hoped that all high-level laws could ultimately be derived from—that is, explained by—fully general microphysical laws. This was supposed to vindicate the high-level sciences by showing their basis in microphysics. To Oppenheim and Putnam, the only alternative seemed to be acknowledging nonphysical entities like an élan vital or nonphysical soul, a repellent proposition for any sort of physicalist (see Potochnik 2011b). This conception of epistemic vindication via reduction is now out of favor, but I suspect its ghost lingers in the tendency noted above to credit lower-level theories with greater epistemic security than higher-level theories. Yet there is no basis for privileging lower-level theories in this way. There is no reason to expect that, in general, the theories, models, and other scientific products that deal with larger entities and their features are less well supported with evidence than those dealing with smaller entities and their features. Not only are lower-level theories often credited with greater epistemic security, but they are also regularly taken to have more general domains of application. Physics is the ultimate example, for it is described as the only "fundamental" science. Kim (2002) says, as if stating a universally shared view, that "the domain of physics includes all that there is" (16). But here one must carefully distinguish between physical entities and physical theory. Any physicalist will grant that all objects are composed of only physical stuff. In contrast, it is not obvious—and quite likely false—that theories in physics address all phenomena. It would take the success of a strong reductionist program for any future theories of physics to account for sociological phenomena, to choose just one example.

These supposed explanatory and epistemic significances of hierarchical organization arise from a common source, namely, assumptions regarding compositional levels and how lower-level parts and their properties metaphysically determine, or realize, higher-level objects and their properties. The difficulties with these supposed significances also stem from a single source. Most basically, granting the existence of composition and realization is not sufficient support for the idea that scientific theories and other scientific representations also reflect these relationships. Metaphysical determination is a relation or set of relations among properties at different levels. This does not straightforwardly dictate the explanatory or epistemic relationships among the theories and other representations that have been formulated about phenomena at different (so-called) levels. What is more, that these scientific representations are formulated to serve human purposes, cognitive and otherwise, is a positive reason to expect the relationships among them to diverge from the metaphysical relationships among their objects. Our scientific representations are heavily idealized, in ways that reflect our needs and interests. These needs and interests vary across individuals and science. There is simply no basis for asserting that scientific representations are related to one another in metaphysically significant ways. Thus there is no reason to expect priority relationships among the scientific investigations of entities and properties that are ordered compositionally or by realization relationships.

Let's move on to the claims about the significance of levels of organization for relative complexity and rates of change, as described in section 6.1. Both rely on the idea that all changes at a given level are mediated by lower-level causal processes. Higher-level systems are taken to be more complex since they are created via interactions among simpler lower-level elements. It is also supposed that changes occur more slowly at these higher levels, since they are mediated by lower-level changes. Those who defend mechanisms as a way of distinguishing levels articulate a version of this same idea. The conception of nested hierarchies of mechanisms, in which lower-level mechanisms perform roles that contribute to some higher-level mechanism, is one version of the thesis that changes at a given level are mediated by lower-level processes—in this case, mechanisms. But these ideas about the causal significance of levels are flawed, at least when applied to the classic compositional hierarchy.[6] Guttman (1976) argues that, just as an object may have important parts at several different levels of organization, interactions between systems at a given level may be mediated by objects at several different levels. For instance, consider the variety of types of interactions among organisms. One organism may causally influence another via the mediation of pheromones, a type of molecule; may causally produce another via a gamete, a type of cell; or may causally influence another via attack and ingestion, which involves much or all of the whole organism. This is clearly so on, at least, the manipulability approach to causation that I invoked in Chapter 2. A corollary to this point is that causal interactions need not occur only among entities at the same level of organization. Very small objects can causally influence objects taken to be many levels higher: recall the crucial role of waste molecules in an ecosystem, for nutrient circulation and water storage. High-level objects can also causally influence objects at much lower levels, what is sometimes termed "downward causation" (Campbell 1974; Eronen 2013). All of this is unsurprising, given the idea of widespread causal complexity I have motivated.

Causal relationships across levels undermine the idea that all changes are mediated by causal processes solely at a lower level, thereby removing the basis for the claim that complexity always increases as the hierarchy is ascended. And there is indeed reason to doubt that ascending compositional levels are always of increasing complexity. Strevens (2006), for instance, proposes an account of how simple behavior emerges in high-level systems, even when the parts of the systems interact in complex ways. One of his primary examples is the relatively simple behavior of whole ecosystems, as seen in the wide applicability of the Lotka-Volterra equations for predator-prey cycles. There is also no reason to conclude that rates of change always slow as the hierarchy is ascended. Consider the sudden extinction of a large population of organisms or even an entire ecosystem. In contrast, genotypes are exceedingly conservative, and the evolution of a new gene complex can take thousands of generations. The type of change in question is a more significant determiner of rate than is the size of the entities involved. Abandoning the idea that changes are always mediated by lower-level processes undermines the idea that mechanisms provide a way to distinguish among classic levels of organization. The parts and wholes of the compositional hierarchy do not uniformly comprise nested levels of mechanisms. It may be possible to provide an alternative definition of levels in terms of mechanisms (see the next section), but it must be emphasized that this conception of mechanism levels does not cohere to the classic conception of hierarchy. Accordingly, such a conception does not have the implications commonly associated with the traditional levels of organization.

In summary, no account of general, hierarchical levels of organization succeeds, nor do the metaphysical, explanatory, epistemic, or causal significances variously ascribed to levels of organization. The relationship of composition may be straightforward and universal, but discrete, stratified levels among the entities related compositionally are not. Nor are levels distinguishable on the basis of realization relationships. Attention to the notion of causal complexity and the strategy of continuing idealization help to separate the straightforward, plausible claims about composition, size ordering, and so on from the dubious ideas about how targets of scientific investigations relate to one another that these claims have so often inspired. The result of this accounting is a thoroughgoing rejection of stratification into classic levels of organization and many of the philosophical significances this idea has been taken to have.

6.2.2 PRIZING APART FORMS OF STRATIFICATION

I suspect that the great appeal of the notion of levels of organization, and the various significances it has been taken to have, derives from its basis in commonsense or even obvious ideas about the hierarchical ordering of our world. These include part-whole composition, ordering by spatial and temporal scales, and functional specifications that can be accomplished via different means. I have argued against a conception of discrete, stratified levels of organization, but I do not dispute any of these ideas at its root. Further, it may well be possible to develop specific hierarchical orderings that are useful in some areas of science. In this section, I discuss the reasonable ideas that have been used to motivate levels of organization and analyze the similarities and differences among them. I then suggest a more limited approach to hierarchical stratification as an alternative to classical levels of organization, and I introduce two examples of this approach. Nonetheless, I finish by proposing that the general concept of levels of organization should be abandoned in philosophical discourse and perhaps in science as well.

Two of the commonsense ideas that seem to inspire the notion of levels of organization are part-whole composition and scale relationships. The latter includes both spatial and temporal scale. Much confusion is avoided if compositional relationships, temporal scale relationships, and spatial scale relationships are all kept distinct from one another. Craver (2007) and Bechtel (2008) suggest that relationships of spatial scale follow from relationships of composition. On the face of it, this is plausible. Because no part can be larger than the whole it composes, spatial relationships in some ways track compositional relationships. However, as Eronen (2013) points out, different parts of a single whole can vary radically in size. Consider again that the causally significant components of an ecosystem include both whole organisms (indeed, organisms of radically different sizes), as well as molecules of waste.[7] A given type of entity can also be a proper part of wholes with radically different sizes. Some sodium ions are independent parts of elephants, while other sodium ions partly compose saline solutions of many different sizes, ranging from those contained in droppers to our oceans. Eronen also points out that relationships of spatial scale exist even where compositional relationships do not. The observation that a waste molecule is much smaller than some elephant does not demonstrate that the waste molecule is part of that elephant or even that any waste molecule is part of any elephant. So compositional relationships are neither necessary nor sufficient to determine the size relationship between types of entities; these relations vary independently.

Now let us consider possible connections between spatial and temporal scale. That these can vary independently follows from considerations introduced in the previous section. Some very small things undergo certain types of changes incredibly slowly, whereas very large things can undergo other types of changes relatively quickly. For example, creep in an individual solder joint of an old lead water pipe occurs very slowly, but the catastrophic failure of an entire water distribution system resulting from cumulative solder creep is a sudden event. Usually the death of individual cells occurs much more frequently and quickly than death of a multicellular organism. But an organism might die suddenly, and most of its cells will take longer, and some much longer, to die. This point is particularly easy to establish from the perspective of a manipulability approach to causation. The type of causal relationship—from a manipulability point of view, the choice of causal variables—is a better guide to speed of action than is the size of the entities involved.

Those who expect a connection between spatial and temporal scale seem to be influenced by the idea that changes at a given level are mediated by lower-level causal processes. This is particularly plausible if one subscribes to causal foundationalism. Causal foundationalism, introduced in Chapter 2, is the idea that there is a microphysical basis for all causal relationships. If higher-level causal relationships are always microphysically accomplished, then one might think that the relatively lower level is both smaller and quicker than the higher level. But, even granting causal foundationalism for the sake of argument, this conclusion does not hold. Regardless of causal metaphysics, it is clear that humans regularly make relatively "high-level," or abstract, causal assertions.[8] Indeed, most of our causal assertions are of this type; certainly all causal claims about middle-sized objects fall into this category. If such abstract causal assertions trace back to microphysical relationships in any discernible way, they are only very distantly related. Accordingly, we should not expect the relationships among these relatively abstract causal assertions to be ordered by size and speed.

The idea that changes at a given level are mediated by lower-level causal processes might also be motivated by appeal to a mechanistic conception of levels. This approaches the matter from the top down instead of from the bottom up, as causal foundationalism does. Here the idea is that any high-level causal relationship must be accomplished in some way or other. That accomplishing involves other causal relationships, and these must be quicker in virtue of being steps of the focal abstract causal relationship. One might also expect these causal relationships to be among smaller entities, in virtue of them occurring (presumably) among the parts of the focal phenomenon. But mechanistic levels do not conform to any of these other hierarchical concepts we have been investigating. Part-whole composition is not sufficient for what we might call mechanistic composition. A proper part of an entity may play no distinct role in a causal process that entity undergoes. Furthermore, where mechanistic levels are identifiable, they do not generally resemble the classic levels of organization. Causal mechanisms and their components may be ordered by both size and speed. However, as we have seen, speed ordering and size ordering don't generally cohere. And both orderings together are not sufficient to guarantee the component-mechanism relationship. Finally, as Eronen (2013) notes, mechanistic composition does not provide a basis for comparing the size and speed of different components of the mechanism.

Another hierarchical concept is that of realization. Roughly put, the relationship of realization obtains between a property implicitly or explicitly defined by its functional role and the entity or entities that fulfill that role. Realization has largely received attention in virtue of the idea of multiple realization. So, for instance, Fodor (1974) talks about the variety of physical realizations of monetary systems, such as those that include strings of wampum and those that include dollar bills. Note, though, that dollar bills (and wampum) are also defined in terms of their causal roles. Dollar bills are in turn realized by certain pieces of cotton and linen paper. Above I suggested that complex realization undermines any connection between realization relationships and compositional relationships. A long list of properties, possessed by a variety of entities, may realize some functional property. Consider that, for a hunk of stuff to qualify as a dollar bill, that stuff must be paper made of cotton and linen, of the proper size, with the proper marks, made through an acceptable process undertaken by a governmentally sanctioned body. What properties are significant for qualifying something as a dollar bill are not structured according to compositional relationships and are notable only for their significance to the functional property in question (and, perhaps, for other directly related properties).

The realization of functional properties seems to bear some connection to mechanistic composition. Intuitively, we might think of some forms of realization, along with mechanistic composition, as relationships of "accomplishing." One might consider what properties are jointly important to realizing the property of, say, being a corkscrew (an example from Shapiro 2000), or one might consider what entities, in what organizational relationship and performing what activities, comprise the mechanism of a corkscrew. According to Haug (2010), this connection holds between mechanistic composition and instances of realization that depend upon causal powers. Similarly, Polger (2010) suggests that the realization relationship might well be involved in mechanistic composition. But mechanistic composition and realization are not interchangeable concepts, nor do they covary. Some forms of realization seem not to involve accomplishing—that is, may not hold in virtue of causal powers. Piccinini and Maley (2014) explicitly define realization in terms of causal powers, but even if they are right, realization seems not to suffice for mechanistic composition. Some realizers are not naturally construed as mechanisms, that is, as integrated entities involved in regular causal processes. This is especially clear in light of complex realization.

There is one final hierarchy concept I will address. Above I identified a sense of "high level" that refers to abstractness: a representation may be considered high level in virtue of omitting reference to many details of the phenomenon that is represented. It seems this best captures the comparison made in Putnam's (1975) famous example of the advantage of high-level explanation: a high-level geometric explanation for why a square peg won't travel through a round hole is better than a detailed microphysical explanation for the same (or so Putnam asserts). The abstractness of a representation can seem related to part-whole composition and indeed to all the other hierarchical concepts I have discussed so far. If you represent only big things, long time periods, and general functional relationships, you omit lots of details. But you also omit many details when you represent only immediate interactions among microphysical entities—namely, how those relate to big things, long time periods, and general functional relationships. To assume information about the latter is somehow implicit in the former is to pretend our microphysics is something other than it is. And, at least on the view of science motivated in this book, there are also principled reasons why our microphysics will never develop to do this. Moreover, abstractness is a property of representations, not of the world. There are more or less abstract representations of the microphysical, more or less abstract representations of the biological, and so on.

Table 6.1 summarizes the hierarchical concepts I have identified, as well as the relations I have suggested hold—and don't hold—among them. The independence of each of these hierarchical relationships further erodes the basis for well-founded, general levels of organization. There is merely a range of hierarchical relationships that partly correspond to one another but mostly diverge. The differences among these relationships inspire a number of the criticisms of levels and the significance of levels I raised in the previous section.

All of this suggests an alternative approach to thinking about hierarchical relationships. Instead of asserting universal levels that have significance for these various hierarchical concepts, one might instead choose a single concept as the basis for the articulation of nested hierarchical relationships. I will discuss two examples of this approach. One is a conception of levels based on mechanistic composition, and the second is a conception of what Potochnik and McGill (2012) call quasi-levels based on spatial scale. Note at

TABLE 6.1. These hierarchical concepts are often assumed to cohere, but I have argued that they are separate. This additionally undermines the idea of universal levels of organization.

Hierarchical concept	Relationships with other concepts
Part-whole composition	Spatial ordering of wholes and their proper parts
	No spatial/temporal comparison of parts or across wholes
Spatial and temporal scale	Vary independently
	Do not require or follow from compositional relationships
Mechanistic composition	Distinct from part-whole composition
	Spatial and temporal ordering of components and mechanisms
	No spatial/temporal comparison of components or across mechanisms
Realization	Does not follow compositional relationships
	Can be involved in mechanistic composition
Abstractness	Representational, not metaphysical
	Fully distinct from all of the above concepts

the outset, though, that there are two limitations to this kind of approach. First, we cannot expect any resulting hierarchy to have direct significance for any of the other types of hierarchical relationships. Second, a hierarchy that results from this approach will only be as scientifically important as the concept on which the hierarchy is based. Even then, a hierarchical ordering itself may be more significant than articulating levels based on that ordering. We may well find that some hierarchical orderings are useful in some specific domains of science but not others and, perhaps, that other hierarchical orderings are of no scientific use whatsoever.

One example of articulating levels based on a specific hierarchical ordering proceeds from mechanistic composition, where a level is defined as all the components of some mechanism. Each of these components may itself be a mechanism with components of its own—those are at a lower level. Sometimes mechanistic levels are discussed as if they are supposed to conform to the classic levels of organization. This might be inferred, for instance, from Craver and Bechtel's (2007) comment that "levels of mechanisms are a species of compositional, or part-whole, relations" (55). But Craver (2007) points out that levels concepts are "multiply ambiguous," and he is clear that mechanistic levels are a special, distinct form of ordering. My analysis of hierarchical concepts also suggests that mechanistic levels do not conform to other concepts. It is better, then, to follow Craver in explicitly developing an alternative hierarchical ordering on the basis of mechanistic composition. For example, in Craver's example of spatial memory, he distinguishes among the following levels, in decreasing order: (1) spatial memory, (2) relevant properties of various brain regions, (3) relevant actions of neurons and synapses, and (4) relevant neural receptors and ions.

The question, then, is how useful this conception of levels will be and in what regards. We can get some initial insight into the potential usefulness of mechanistic levels from the paradigmatic scientific research on which philosophical treatments of mechanisms focus. These include, prominently, focal phenomena in neuroscience and molecular biology, like neurotransmission and DNA replication, respectively (Machamer et al. 2000), as well as spatial memory, which Craver and Bechtel (2007) and Craver (2007) discuss. From the perspective adopted in this book, it seems that when a causal pattern involves many steps that regularly occur together, in a specific order, and by several components of a somewhat integrated body, the analogy with human-designed mechanisms is apt, and mechanistic composition may have scientific significance. Add to this a research focus on the integration and distinct causal steps, as is common in neuroscience and molecular biology, and mechanistic thinking is of clear scientific value. I suspect that mechanistic thinking is of little use when either the nature of the causal relationships or the research interests diverge from this rough characterization (see also Levy 2014). For example, the sense of "mechanism" is strained at best when it is applied to causal relationships like that between natural selection and evolved traits (Skipper and Millstein 2005).

So now we know some of the kinds of scientific projects that may benefit from emphasizing mechanistic composition. But even in those cases, a conception of mechanistic *levels* is not very successful. As Eronen (2013) points out, mechanistic composition does not enable levels to be distinguished across different mechanisms. This has a counterintuitive consequence: different subcomponents of a mechanism cannot be grouped into a single level if their contribution proceeds via different components of the mechanism. In the example of spatial memory mentioned above, a neuron in the hippocampus is not on the same level as a neuron in the prefrontal cortex, even if both contribute to spatial memory. These neurons are not on different levels either but incomparable ones. Further, Haug (2010) points out that the same entities can serve as components of different mechanisms. Taking into account the earlier point, this means that one entity that participates in multiple mechanisms is itself located at incomparable levels. Once mechanistic composition is fully divorced from part-whole composition, it is impossible to compare levels of components across different mechanisms or to compare levels of subcomponents across different mechanism components, even if the components or subcomponents are similar or even the very same object. A schema of this ordering is shown in Figure 6.2. At best, there can only

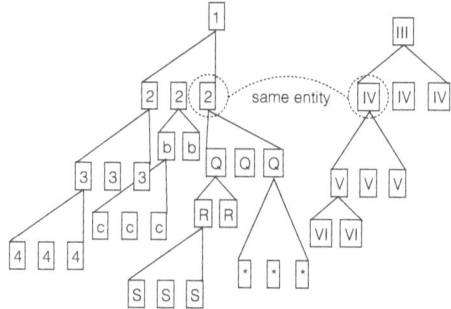

FIGURE 6.2. Mechanistic "levels" are more accurately described as incomparable nesting relationships. Levels can only be identified within a mechanism (or submechanism), and the same entity may be active in different mechanisms and thus at incomparable levels.

be radically contextual mechanistic "levels." But the concept of levels does not seem apt; such relationships seem to be more accurately described as mechanistic nesting relationships.

Division into mechanistic levels is thus extremely limited and perhaps undeserving of the term "levels." That division is also variable and dependent on the immediate research interests. Highly similar entities, like neurons, or even the very same entities, like a single neuron, that are involved in multiple mechanisms or multiple components of a single mechanism will be judged to be on different levels depending on the task at hand. One result of all of these limitations is the highly circumscribed relationship between mechanistic levels and spatial and temporal ordering I introduced above (Eronen 2013). Spatial and temporal scale comparisons can be made between a mechanism and one of its components. However, no conclusions can be drawn about the similar scale of parts of a single mechanism (the identically numbered or lettered elements in Figure 6.2) or about scale similarities or differences across mechanisms or submechanisms. All of this strongly suggests that mechanistic levels may not be of much scientific significance even when mechanistic composition is.

Another approach to defining levels explicitly based on a particular hierarchy concept is developed by Potochnik and McGill (2012). In this instance, the hierarchy concept in question is spatial scale. Brian McGill (2010) shows how explicit consideration of scale, particularly what causal influences dominate at different scales, has proved enlightening in the field of ecology. His main example is how the influences on species distribution change with the size of the area under consideration. Unlike part-whole or mechanistic composition, spatial scale does not in itself provide the basis for discrete intervals. This is provided by the focus on causal patterns. For phenomena that can be analyzed at different spatial scales, such as species distribution, different causal influences may dominate at different scales of analysis. If so, that provides the basis for articulating levels, or what McGill and I would prefer to call "quasi-levels." This terminology indicates the provisionality of the determination of levels. Whether differences in scale are of causal significance depends on both the system under investigation and the specific research question. Divisions into quasi-levels are, then, patently not of universal significance.[9]

Consider a simple example from Potochnik and McGill (2012). A squirrel may stand 0.1 meters high while a tree that it lives in stands 10 meters high. Considered in terms of the traditional part-whole compositional hierarchy, the squirrel and tree are at the same level: the level of individual organisms. Yet their difference in scale is significant for some investigations. Masting occurs when trees produce all their seeds in large bursts, which happens only in some years. For investigations into the ecological sources of the evolution of masting, the salient relationship is between a population of squirrels and an individual tree. One squirrel does not eat enough seeds to drive trees to evolve masting; it takes an entire population. Indeed, the time between masts has evolved to time periods roughly equal to the generation time of a squirrel (about two years) so as to have maximal influence on the population dynamics of the squirrel while minimizing influence on the tree's fitness. In contrast, for investigations of populations' northward progression in response to global climate change, the difference in size between a squirrel and a tree is insignificant. Their dispersal distances are very similar; indeed, there is evidence that the rate at which oak populations spread is heavily dependent on the dispersal of seeds by squirrels (Corre et al. 1997; Clark 1998). In this example, scale and part-whole composition part ways, as anticipated. Furthermore, the difference in scale between the squirrel and the tree is important in one scenario (masting evolution) but not in another (dispersal).

At root, the concept of quasi-levels is grounded in the recognition of the scale-dependence of many causal patterns, along with the general observations of causal complexity and interest-relative representation. It is different from the other hierarchy concepts I have discussed in some important ways. Division into scale-based quasi-levels is a form of hierarchical organization that explicitly depends on both the nature of causal patterns and the interests of scientists. The determination of what scale range to group as a single quasi-level depends on what causal patterns are focal, as well as on how those focal causal patterns behave: what entities or groups of entities interact, at what scale different causal processes become dominant, and so forth. This relativity to research interests and the nature of causal patterns makes this approach to defining levels at once very broadly applicable and, at the same time, not very informative. Applying the concept of quasi-levels does not answer the question of the significance of size ordering; instead, it is simply a way to explicitly raise that question. For this reason, searching for quasi-levels in the context of some particular scientific investigation makes salient the possibility of unexpected types and directions of causal influence, possibilities that intuitive levels concepts may lead scientists to overlook. This is a notable difference between quasi-levels and all other hierarchy concepts I have surveyed. Further, because of the concept's limitations, there is no expectation that the successful demarcation of quasi-levels has ontological significance, or even significance for related phenomena. The question of whether there are discrete levels and, if so, how to articulate them arises anew with each investigation. The flip side of this limitation of quasi-levels is its reapplicability. We can expect many different divisions into quasi-levels based on how different phenomena respond to changes in spatial scale, and for some phenomena, quasi-levels might instead be articulated based on temporal scale. As discussed above, temporal scale cannot be assumed to correlate to spatial scale. Accordingly, one must always choose a single basis for articulating quasi-levels for a given research focus.

The critical view of levels developed in this section suggests that any successful conception of scientific levels will be significantly limited in its implications, including its implications pertaining to other hierarchy concepts. A successful approach to distinguishing levels will quite possibly also be domain and interest relative or may only be successful in some limited domain. Accordingly, I close this section with a proposal for philosophers and perhaps for scientists as well. Let's jettison talk of levels entirely. If what I have said in this section is right, then it is possible to continue to employ the concept of levels, so long as one is careful to articulate what exactly is—and is not—intended by any given reference to levels. But, to appeal to Eronen (2013) once more, "the danger in this is that other intuitions about levels may creep in" (1049). These various intuitions about levels are, after all, the basis for the standard picture of a stratified reality that I have called into question here. Of the concept of quasi-levels, I have stressed that it highlights the variability of causal relationships—that it introduces questions about the significance of scale rather than purporting to answer them. Ultimately, perhaps the most important feature of this concept is the adjective "quasi," advertising its limitations.

6.3 The Fields of Science and How They Relate

Classically, a neat alignment has been assumed among compositional relationships, levels of organization, and the division of scientific labor. I have pointed out that claims about the significance of levels regularly relate metaphysical levels of organization to levels of scientific investigation. The idea of an organizational pyramid has been common in both philosophy and science and, as that metaphor suggests, levels of scientific investigation and of entities are classically taken to conform to the fields of science with which we are familiar. The relationship of this posited organizational scheme to the division of scientific labor is part of what is at stake in the philosophical debate about the unity or disunity of science. The classic positions on that debate are represented by Oppenheim and Putnam (1958) on one hand, who hypothesize that all fields of science will eventually reduce to microphysics, thus creating a unified science, and Fodor (1974) on the other, who argues that economics and other so-called special sciences do not reduce in this way, and so science is disunified.

A science without discrete, universal levels of reality to investigate interferes with this conception of the field divisions of science, and it challenges both the classic positions of unity and disunity of science. Here I propose an alternative account of the division of science into distinct fields and of the relationships among those fields. The same reasons we have for thinking that our scientific representations are not hierarchically ordered also cast doubt on the idea that our fields of investigation are ordered in this way. As I have explored in earlier chapters, causal complexity necessitates different idealized representations to accomplish different scientific aims. There is no reason to expect these idealized representations to track the complex and variable facts about composition relationships or realization relationships. Indeed, I have argued that causal relationships are not reliably structured by compositional or realization relationships. On the view of field divisions in science this inspires, our scientific fields have little in the way of metaphysical significance. Instead, their origin is merely in their usefulness to humans with various aims of understanding and control.

It may be possible to partially order some of our scientific fields according to the size of the entities of primary interest in those fields, and this might adequately capture some of the intent behind the classic characterization of field divisions. So let's investigate the prospects of that approach. Notice first that this does not work particularly well for the broadest classification scheme: physics investigates the infinitesimal as well as the entire universe, and biology's range is only somewhat narrower, encompassing both RNA and ecosystems. Nor is it possible to arrange these fields according to the breadth of the types of objects they investigate, with physics as the most general. Physical theory and models are of little relevance to ecosystems, and the conception of the aims of science I have articulated provides reasons to think that won't change. Midrange classification schemes may fare the best in ordering scientific fields according to the size of objects under investigation, but the results are somewhat nontraditional. Microphysics deals with the very small, molecular biology the somewhat less small, and physical chemistry the larger still. Physiology tends to deal with the medium-sized (defined as very roughly human-sized), population biology larger entities, geology the very large, and astronomy the enormous. Finer field divisions render size ordering impossible and also cast doubt on the midrange classification I just suggested. For then biogeophysics, cosmology, and multidisciplinary investigations of circadian rhythms arrive on the scene, bringing together the very large and the very small and thereby wholly undermining any potential ordering of fields by size of objects under investigation.

There is only a very limited and obvious sense in which our broadest fields of science correlate with the status of the entities they investigate in a way consistent with classical treatments of levels of organization. This sense is reflected in our terminology: sociology regards social interactions, biology biological phenomena, chemistry chemical processes, and so on. And then, social interactions tend to occur among biological entities whereas biological entities need not be part of any social structure, biological phenomena involve some but not all chemical processes, and so on. But even here the limitations are more impressive than the observation of a general relationship. First, these namesakes may be the primary focuses of the fields investigating them, but they are not the exclusive focuses. As I discuss in greater depth below, cross-connections among investigations in different fields are regular and important. In the course of investigating focal phenomena, scientists regularly have need to consider types of phenomena that are of only derivative interest, some of which are primarily targeted in wholly different fields of science. For instance, social context can be essential to accounting for some phenomena involving biological individuals, like traits that result from frequency-dependent selection. This too is due to causal complexity. Second, the focal phenomena in these fields need not be nested. Social phenomena, for example, might exist of aggregates other than biological entities—candidates include networked computers and business entities that do not depend on any specific biological individuals.

The failure of our fields of science to correspond to levels of reality is another wedge between scientific practice and metaphysical implications. This follows the pattern we have already seen of a feature of the scientific enterprise reflecting back to us our interests and needs in a way that interferes with a transparent portrayal of the world. Science is divided into fields (as well as subfields and sub-subfields), and these divisions persist for reasons that are by this point familiar. The world is complex. Our interests, cognitive resources, and other resources are limited. Groupings into fields, subfields, and other units result from commonalities in focus, methods, or background concerns. Thus, the divisions in our human-built science partly reflect the order of the world, and they partly reflect the nature of human interests. They also just as surely reflect historical accident and contingencies of social organization (Adrian Currie, in conversation).

Let's move on to another issue, namely, what can be said on the present view of the relationship among different fields of science. The conception of hierarchically ordered fields of science is, as I mentioned, presumed in both of the classic positions on the debate over the unity or disunity of science. Understood in that way, the declaration that science is or will be unified is the thesis of explanatory reductionism, according to which the unity of science is achieved by explaining all laws of science in terms of their connection to microphysical law. This is the explanatory and epistemic significance for levels of organization proposed by Oppenheim and Putnam (1958), among others. The declaration that science is and will remain disunified is, in turn, the rejection of the thesis of explanatory reductionism. A proponent of disunity may hold that there are different explanations appropriate to different "levels" (Fodor 1974; Dupré 1993) or that the findings from different fields are not neatly related and cannot be combined (Cartwright 1999). The latter version of the disunity thesis is the only one among these positions that does not presume the hierarchical ordering of fields of science.

There is, however, a somewhat neglected tradition that offers a wholly different conception of the unity of science. This alternative conception of unity does not require hierarchically ordered fields, but it nonetheless has the potential to provide some of the advantages promised by advocates of the more traditional unity based on the explanatory and epistemic reduction of high-level sciences to the fundamental level of physics. According to this alternative tradition, the unity of science is achieved via the coordination of diverse fields of science, none of which is taken to have a privileged epistemic status. This idea has roots in Otto Neurath's notion of unified science, and it has been developed in different ways by some philosophers in the intervening century. I have termed this family of views "coordinate unity," in contrast to what one might call reductive unity (Potochnik 2011b).

Among the members of the Vienna Circle, Neurath was the most active proponent of the unity of science. Prominent advocates of explanatory reductionism—Oppenheim and Putnam (1958), Nagel (1961), and Hempel (1966)—were directly influenced by the logical empiricism of the Vienna Circle, but their reductive approach to unity bears little resemblance to Neurath's view. Instead, Neurath's vision of unified science is, at root, a practical aim: bringing together scientists in distinct fields of science to facilitate communication and interconnection. He thinks that a prior determination of the forms of interconnection would be impossible. As Neurath (1937) puts the point, "Instead of aiming at a synthesis of the different sciences on the basis of a prior and independent philosophy, the special sciences will themselves supply their own synthesizing glue" (172). Because there is no preconceived notion of the ways in which the different parts of science will fit together, unified science will not create a complete, all-encompassing system. Instead, it is an ongoing process of systematization, fostered by collaboration among those engaged in diverse scientific endeavors. The result is simply "a preliminary assemblage of knowledge ... the totality of scientific matter now at our disposal" (Neurath 1936a, 146).

In Neurath's view, the need for these forms of collaboration arises in the process of applying science to particular problems. He writes,

> We avoid pseudo-problems of all kinds if, in the analysis of sciences, we set out from predictions, their formulation and their control. But it is precisely this starting point that is little suited for the delimitation of special disciplines. One does not arrive at individual disciplines of stars, stones, plants, animals during the deduction of certain predictions, because time and again the conjunction of statements of different origin become necessary. (Neurath 1936b, 132)

Our scientific aims regularly force upon us unexpected connections among different parts of science. Predicting and accounting for phenomena in the natural and social world regularly require that distinct scientific disciplines work in conjunction. This element of Neurath's view is, to my mind, crucial. It indicates a form of interconnection that is at the very heart of science. Simply put, evidential relationships do not respect field boundaries. For this reason, accurately treating phenomena of interest regularly requires collaboration among different fields of science, despite divergences in their domains of inquiry, primary aims, methods, and resulting details of execution.

Consider an example from evolutionary biology. It's widely presumed that the peacock's colorful train evolved in response to female peahen mating preferences. Takahashi et al. (2008) investigated this proposed explanation. They observed a population of peafowl for several years; their observations suggest that the length and elaborateness of trains actually do not affect peacocks' mating success. This approach is typical to field ecology, but the study does not end there. Takahashi et al. also cite findings from phylogeny and endocrinology that support their conclusion. They note that these features of peacock trains have been found to be under estrogen control (Owens and Short 1995) and that male plumage under estrogen control has been found to be disregarded in mate choice. They also cite a molecular phylogeny finding that suggests that all peafowl had bright tail plumage in the evolutionary past (Kimball et al. 2001). This suggests that, instead of peacocks evolving a more elaborate train, peahens actually evolved a less ornamental appearance. These findings further undermine the idea that the peacock's colorful train evolved because of peahen mate choice.

Takahashi et al. (2008) define their investigation not by methodology or by field of research but by a particular phenomenon of interest and a focal causal pattern: the existence of the peacock's colorful train and what role, if any, mate choice played in its evolution. They take into account findings from field ecology, phylogeny, and endocrinology. Their focal causal pattern leads them to emphasize the train's influence on mating opportunities, which is thought to be a source of selective advantage. But other causal influences on the peacock's train are also taken into account, though only as a way of evaluating whether the focal causal pattern is embodied. Evidence marshaled from endocrinology provides insight into the developmental causes of the colorful train, and these limit the possible targets of selection. Molecular phylogeny provides information about the evolutionary past, which limits the possible evolutionary trajectories. From the phylogenetic evidence, it emerged that the proper question is not what caused peacocks to evolve more colorful trains in contrast to peahens but what caused peahens to evolve duller coloring and smaller trains from an ancestral state of colorful, long trains. The focal causal pattern is apparently not embodied by this phenomenon. This may suggest limitations of causal patterns involving female mate choice's influence on ornamentation more generally, as the peacock's train is taken to be emblematic of that form of sexual selection (Roughgarden 2009).

The regular need to look to other fields and subfields for evidence can be used as the basis for a promising conception of the coordinate unity of science. I call this conception of unity "epistemic interdependence". In earlier chapters, I have argued that causal complexity motivates a diversity of scientific aims and methods, as well as idealized representations particular to their purpose. In many ways, this is an account of science as disunified, as in-principle unintegrated. But grappling with causal complexity also regularly requires some forms of collaboration across these diverse projects, including even across different fields of science. As illustrated by Takahashi et al.'s (2008) research on the peacock's colorful train, evidential relationships cross field boundaries in a way that can motivate collaboration. The resources of separate fields and subfields regularly must be brought together to provide the evidence needed for some research program to accomplish its specific, limited aim. This is epistemic interdependence. It constitutes a meaningful sense in which science is a unified whole, even as scientific representations and the aims they serve remain separate and unintegrated.

The nature and extent of collaboration in service of epistemic interdependence are variable, for different kinds of collaboration can contribute to evidence gathering across research programs and field boundaries. But whatever form connections among different research programs and fields take on, these never interfere with the persistence of idealized representations in service of specific, limited aims. Indeed, as in the research on the peacock's train, often this evidence gathering serves to corroborate or undermine the acceptability of idealizations. Here I discuss three possible forms of collaboration in service of epistemic interdependence. The first two are frequently employed, whereas the third is only pursued in limited circumstances. Those limitations are themselves important to my conception of coordinate unity as epistemic interdependence.

First and most obviously, marshaling evidence from diverse sources frequently involves sorting out the type and direction of causal influences, even for causal factors that are not of primary research interest and are, thus, idealized. This is important for many scientific aims. It's useful for uncovering general causal patterns and for determining whether and to what extent particular phenomena embody those patterns with the aim of generating understanding. It is also important for establishing whether a predictive model is sufficiently accurate. This kind of causal analysis across research programs occurs in Takahashi et al.'s (2008) investigation of the influence of mate choice on the evolution of the peacock's colorful train. The researchers appeal to a finding from endocrinology that peacock trains are under estrogen control. This causal influence is incidental to the researchers' main focus, except as it bears on the possibility that the trains evolved in response to female mating preferences. Establishing the relevance of the endocrinological finding of estrogen control for the sexual selection hypothesis thus also requires evaluating causal relationships between developmental processes and selective influence. In particular, the finding that male plumage under estrogen control is typically disregarded in bird mate choice is significant in this context. The finding that the peacock's train is under estrogen control thus suggests that this phenomenon is unlikely to embody the focal causal pattern of having evolved due to female mate choice; Takahashi et al.'s field study is a step toward confirming this. This illustrates how causal investigation can benefit from collaboration or at least the sharing of results among different labs, even in different subfields or fields of science.

Two limitations of this form of collaboration across research programs are important to note. First, causal analysis across research programs to generate evidence generally occurs in service to preexisting research aims. I expect that collaboration will only infrequently include the development of shared research aims or methods or integrated representations (see below). Second, marshaling evidence regarding causal influence is only useful to the extent that it helps determine whether a posit meets the applicable standard of acceptability. For the aim of understanding—or, equivalently, explanation—I have argued that the relevant standard is epistemic acceptability. Recall that epistemic acceptability requires a different degree of accuracy of posits, depending on their role in a representation and the purpose to which the representation is put. In only special cases does this require a posit to be true. Causal analysis across research programs thus must proceed in a way that is sensitive to whether truth or accuracy in some regard is required for acceptability, epistemic or otherwise, given the specific aim of the research. Demonstrating that an idealization is wildly inaccurate is not useful so long as the idealization still adequately serves the limited research aim at hand. Parameters should check out as accurate enough for the purpose they serve. Causal factors that would threaten to obliterate a posited causal pattern, such as estrogen control of male plumage suspected to result from sexual selection, should be absent or taken into account.

Research programs' different aims and the different idealizations that serve those aims often necessitate a second type of collaboration to discern evidential relationships that cross field boundaries. This is collaboration to navigate the differences introduced by distinct research aims. Such differences can include differences in terminology and categorizations, incompatible idealizations and other posits, and different applicable forms of acceptance for posits, even apparently similar posits. Consider terminological differences. The terminology of different research programs can diverge even if their investigations are related, and differences in terminology can block the way to effective evidence sharing. Locating overlaps and discrepancies in the terminology of different investigations can help identify the nature of causal relationships with relevance that spans those investigations and can help prevent mistaken inferences. The use of the term "gene" in different contexts is a nice example of this. The clearest divergence in the concept of a gene occurs between classical genetics and molecular genetics, but some argue that the gene concept varies even within these theoretical contexts (Kitcher 1992; Dupré 1993; Rosenberg 1994). Different gene concepts, serving different research aims, lead to different, apparently incompatible causal claims. For instance, the relative causal importance of natural selection and genetic drift can depend on whether the gene is understood in the classical sense or the molecular sense. We have seen many examples of different, incompatible posits that serve different aims: representing a population as infinite or as a specific, nonactual "effective" size; idealizing away evolutionary influences besides natural selection or focusing on one or more of them; and so on. Successfully identifying and navigating such differences is necessary to correctly gauge how findings in different research programs relate.

Let us consider a third, more ambitious step toward marshaling evidence that transcends field boundaries, namely, the development of integrated representations—models or other representations of the interplay among causal dynamics that had been treated separately. I expect that this strategy will be employed only in conditions similar to those that, in Chapter 5, I suggested warrant integrated explanations: either when a specific phenomenon is of particular scientific interest or when the causal interplay in question recurs frequently enough to make it a useful focus. There I introduced the example of Roughgarden's integrated approach to studying the environmental mediation of gene expression and environmental sources of selection in the production of social behavior in animals. Roughgarden (2009) points out that many behaviors, such as a bird's allocation of effort to foraging versus nest guarding, are the product of selection working on a set of conditional strategies. She consequently suggests developing integrated models that simultaneously represent selection pressures and the conditional strategies that influence the behaviors animals actually perform. In Chapter 5, I emphasized the role this can play as an integrated explanation, distinct from but not supplanting existing evolutionary explanations. Integrated models like these, if developed, can also provide a framework for sorting out causal relationships and evaluating evidential relevance. For example, the integrated approach that Roughgarden advocates may show that some particular game theory explanation gets the causal pattern of the ecological sources of selection wrong in virtue of its idealizations failing to accommodate the role of, for instance, learning or development.

Developing representations that systematically integrate diverse causal influences can be a useful form of coordination among different research programs, but it is not often worth the investment. This strategy requires overcoming a host of difficulties and inhibits rather than contributes to the success of many scientific aims. If causes interact in different ways in different specific cases, which is common in our causally complex world, then integrated models will be highly specific and not widely applicable (Mitchell 2003). They will fail to represent the broad causal patterns that I have argued are so crucial to generating scientific understanding. On a more practical level, different approaches tend to resist integration due to incompatibilities like those discussed just above. Pervasive, distinct idealizations and other representational differences are not easily combined or replaced. For instance, population genetic models of evolution represent evolutionary change generation by generation, whereas many game theory models are equilibrium models that represent only long-term evolutionary outcomes. This difference reflects these approaches' different aims. Phenotypic evolutionary change behaves predictably only in the long term, whereas the role of individual genes is most easily discerned in generational change (Eshel and Feldman 2001; Godfrey-Smith and Wilkins 2008). Different idealizations facilitate these different styles of representation. It is difficult to imagine how they would be effectively integrated (James Griesemer, in conversation), and in any case, doing so would inhibit both representations' aims. And so, while the integration of approaches can help sort out evidential relationships that cross field boundaries, this will not often prove to be possible or worth the investment.

This view of the unity of science as epistemic interdependence is a member of the family of views I have termed "coordinate unity". Other views in this family have been developed by Darden and Maull (1977), Bechtel (1984), Wylie (1999), Mitchell (2003), Grantham (2004), and, specifically for neuroscience, Craver and Darden (2005). I do not want to overstate the differences between my view and what I take to be kindred views, but I should briefly indicate what is distinctive—and desirable—about my articulation of coordinate unity. In my view, the coordination of different fields in science arises wholly from the different aims and different deployments of aims that define different research projects, together with the causally complex world that warrants this piecemeal approach to science. This situation, which could be described as the *dis*unity of scientific products, as well as the rampant and unchecked idealization that it motivates, is what gives rise to epistemic interdependence. Epistemic interdependence is simply the existence of evidential interrelationships among different research programs, even across different fields of science. Capitalizing on those evidential interrelationships is the sense in which, in my view, science is unified.

My view of the unity of science thus focuses exclusively on forms of collaboration that contribute to evidence sharing across different projects and fields in the face of causal complexity. This is a narrower conception of unity than the range of interfield connections that Darden and Maull (1977) and Grantham (2004) advocate, and it is broader than the focus on mechanisms that Bechtel (1984) and Craver and Darden (2005) urge. Additionally, the present view differs from many versions of coordinate unity in advocating not interfield theories but piecemeal integration of the sort Neurath envisioned.

My view bears the most resemblance to that of Alison Wylie (1999), who emphasizes the importance of evidence sharing among otherwise disunified fields, appealing to the example of archeology. Perhaps the only difference is that Wylie leaves the door open for unification of theory, which is at odds with my prediction of the persistence of wholly independent scientific aims and products. Finally, although my view also has much in common with Mitchell's (2003), it should be clear from my account of explanation in Chapter 5 and the argument for the separate pursuit of science's aims in Chapter 4 that I do not expect unity across fields to furnish explanations or directly fulfill any other aims. The scientific products that serve different research programs are steadfastly independent. In my view, coordination across projects and fields is only a way to secure the minimal amount of evidential security needed for each research program to pursue its own distinct aim.

This conception of the unity of science is wholly different from reductive approaches to unity, and the differences align with the arguments against the existence and significance of levels in the first part of this chapter. Notably absent from my account of coordinate unity is any reliance on ontological levels of organization or the related expectation of asymmetric relationships among fields. Causal influence can run in any direction, among phenomena dealt with in any field. Accordingly, this conception of coordinate unity does not take some fields of science to be more epistemically privileged than others. Instead, a holistic epistemic security is obtained via a hodgepodge of connections among fields that spring up in response to evidential interrelationships resulting from causal complexity. The same features of the world that undermine the concept of well-defined levels necessitate coordination among fields. Unlike reductive unity, this coordinate unity is not threatened by antireductionists' observations of multiple realization, disparities in the classification of kinds, or different forms of idealization. To the contrary, each of these complicates science such that forms of coordination like those I have outlined are even more essential.

And yet, despite these deep differences, this conception of coordinate unity can achieve one of classic explanatory reductionism's primary aims, namely, greater epistemic security for science. As we have seen, the reductionist approach to epistemic security is to await corroboration of scientific results by our best theories in physics. But current physics is ill-equipped to adjudicate almost any of the findings in other fields of science. If my view of widespread idealization in service of different aims is right, we should not expect this to change. Our physics simply isn't designed to give insight into, say, sociological theory. My conception of coordinate unity as epistemic interdependence offers a wholly different approach to improved epistemic security for the sciences. Systems of mutual revision and reinforcement created by the sharing of evidence are incredibly important to scientific success. Different research programs differ in their aims, their idealizations, and the tools they bring to bear. Yet the phenomena they investigate are not isolated events but different parts of a complex causal web. Because of their different aims and methods, different research programs are well positioned to furnish evidence that others benefit from but cannot easily generate themselves. This coordinate unity of science strengthens the epistemic positioning of all fields of science. This epistemic security via interdependence is multidirectional: evidence can come from any direction, from unanticipated sources, and evidential relationships are often mutually beneficial to the research programs involved.

As an illustration of an unlikely collaboration that leads to mutual epistemic gain, consider the emerging area of investigation called biogeophysics. Biogeophysics examines the causal relationships between microorganisms in the ground and geological features of the Earth. Researchers in the fields of microbiology and geophysics have shown that microbial activity influences subsurface geological features and, conversely, that geological properties influence microbial activity (Atekwana et al. 2000). These findings were initially surprising; some were resistant to the idea that such disparate types of phenomena—one tiny and quick acting and the other enormous and slow to change—could influence one another (Estella Atekwana, in conversation). The key to both is conductivity: electron transport is important to microbes, and conductivity is an important geological influence. The discovery of this causal interrelationship has led to breakthroughs for both microbiology and geophysics. It demonstrates that microbial features may need to be taken into account when investigating geophysical properties and that understanding microbial activity can involve considering the geophysical properties of the ground that houses the microbes. These discoveries also create new research tools. For instance, employing traditional geophysical techniques creates a new opportunity to explore microbial activity noninvasively and in the field (Atekwana, in conversation).

In summary, there is a coordinate unity of science consisting in epistemic interdependence among research programs with different aims and targets of study, regardless of the field that houses them. This unity transcends differences among research projects, subfields, and fields. Causally complex phenomena investigated by different research programs create the need for collaboration. In particular, research programs benefit from considering the evidential implications of related investigations, even if those investigations occur in different fields, invoke different terms or characterizations, employ different idealizations, and pursue different aims. It is in virtue of this collaboration, with the goal of establishing evidential implications in the face of causal complexity, that science comprises a unified enterprise. That unity is not in virtue of reduction to a single, fully general science, nor is it in virtue of similarities among different fields. Instead, the unity of science emerges from the cross-connections that are established—and continue to be established—among related investigations, in a piecemeal fashion, despite differences in terminology, concepts, idealized representations, methods, and aims. These cross-connections are required for evidence gathering to discern causal patterns in our causally complex world. This is a thesis of the unity of science insofar as it (a) posits connections among diverse scientific projects and fields, (b) for a single reason, and (c) in a way that achieves epistemic security—the inspiration for reductive theories of the unity of science.

Causal complexity and a science that serves humans' cognitive and practical needs lead to a variety of fields and subfields of science, with different aims and approaches both across and within those fields. This variety does not correspond to discrete levels of investigation, nor does it isolate entities and properties at discrete levels. But nor is the result incommensurable, isolated scientific endeavors. A meaningful unity across science exists in the shared requirement of evidence gathering in the face of causal complexity and idealized representation and in the collaborations this necessitates. This recalls Neurath's admonishment that we should start out from what is practically necessary for the success of science, always heeding actual scientific practice.

7

Scientific Pluralism and Its Limits

The account of the aims of science I have developed in this book explicitly links those aims to particularities about human scientists and human consumers of science. In Chapter 4, I motivated an alternative construal of science's epistemic aim as not truth but understanding, which enables idealizations to directly contribute to science's epistemic success. I argued that what satisfies the aim of understanding depends on researchers' characteristics and interests, but that this does not result in a problematically subjective standard. In the same chapter, I also suggested that there are many different aims of science and that these diverse aims are best satisfied by different scientific products. There is an overarching theme to these many scientific aims. All further cognition or action; indeed, they further *human* cognition or action, as a science by and for human beings should. In Chapter 5, I developed an account of scientific explanation that is relativized to human explainers in much the same way. And then, in Chapter 6, I argued that the diverse and limited aims of different scientific research programs, and the idealized representations that serve these aims, result in an epistemic interdependence among science's products. This requires coordination among different research programs to the end of sharing evidence across those projects.

In this final chapter, I investigate some implications of these deep connections between science's aims and human particularities. In section 7.1, I consider how this human-centric conception of the aims of science expands the role of social values in science. Then, in section 7.2, I suggest that the human-centric nature of science greatly constrains the potential for drawing metaphysical conclusions from our best scientific findings. This has implications for naturalized approaches to metaphysics. Finally, in section 7.3, I address the concern that the view of science I have presented in this book is too unhinged from realistic depictions of reality to fulfill the epistemic and social functions for which science was designed. I argue that the ideas I have motivated ground a pluralistic view of science but that they also create natural brakes on that pluralism. This discussion inspires a handful of normative conclusions about scientific practice and the direction of scientific progress. Together, these ideas comprise a view of science as responsive to and reflective of human particularities—but in ways that contribute to, rather than undermine, science's success.

7.1 The Entrenchment of Social Values

Linking the aims of science to particularities of scientists and their audiences introduces the possibility that specific values held by individuals influence the products of science. In this section, I develop the details of how this plays out, in light of the views I have motivated about idealization and the aims of science. The basic idea is that rampant and unchecked idealization, motivated by diverse scientific aims, creates inroads for social values to significantly influence the products of science. This conception of science enables simple ways in which particular humans and their values influence science to have great significance, including direct epistemic significance. Rather obvious and mundane ways in which social values influence science turn out to have direct epistemic significance by shaping what is required for scientific understanding. The diversity of science's aims, as well as the diverse scientific products that best serve those aims, also extends the influence of social values on science. Because social values help determine which aim—indeed, which precise deployment of an aim—is pursued, they influence what features our scientific products should have.

Consider the first idea, namely, that social values have direct epistemic significance in science. This stems from my account of the epistemic aim of science as not truth but understanding. As Elliott and Willmes (2013) emphasize, values do not seem relevant to the determination of whether a hypothesis is true. Our values may influence what truths we seek and how we employ those truths, but not what is and is not true. But if science's epistemic aim is instead taken to be understanding, and if I am right that some departures from the truth can promote understanding, this creates additional inroads for the epistemic influence of our values, both shared and individual. Unlike with truth, our values' influence on what we seek to understand and why has direct implications for what does and does not generate understanding.

The epistemic influence of values on science has as its starting point an obvious and rather uncontroversial role for social values in science. This is in the choice of which research questions are pursued and how those research programs are subsequently developed. The potential directions for scientific research are vast. Human concerns and values, in general and of particular scientists, their funders, and so on, shape which of these potential directions for research are in fact pursued. Considerations introduced in this book elevate this mundane role for social values to epistemic significance. Recall that, because of causal complexity, any given phenomenon embodies many causal patterns. The same phenomenon may thus be targeted by many different research programs, in principle or in fact. Different research programs bring different deployments of the aim of understanding (or other aims entirely) to their investigation of the phenomenon. The particular research program that occasions an explanation, with its specific deployment of the aim of understanding, determines which causal pattern explains the focal phenomenon.[1] And so, by influencing which research programs are pursued, individual and shared interests, concerns, and values shape the content of our explanations—not just what we aim to explain but what in fact explains those things. This is one source of social values' direct epistemic significance.

The concerns and values that help guide a research agenda are also relevant to determining the standards for epistemic acceptability and thus the epistemic requirements for legitimate understanding. My proposed requirement for the epistemic acceptability of a posit is that its divergence from the truth be insignificant, taking into account (a) its role in the representation and (b) the epistemic purpose of that representation. Both depend on a research program's particularities. First, the emphasis that different research programs place on different causal patterns results in different choices of idealizations and of elements that are intended to represent realistically (in some regard, to some extent). Different standards govern the epistemic acceptability of posits playing these different roles. Second, research programs emphasizing different causal patterns is simply what it means for them to have different epistemic purposes. The epistemic acceptability of even the same posit will be judged differently when the focus is on different causal patterns. Social values shape the content of our explanations, and they also shape the epistemic requirements that must be met for legitimate understanding, an epistemic achievement.

Consider an illustration of how a research program, and so the concerns and values that help guide it, can have epistemic significance in these ways. To return to an example I have regularly employed in this book, when researchers seek to understand the evolutionary role of natural selection, an optimality or game-theoretic explanation is often called for. These styles of explanation can depict the right kind of causal pattern to generate understanding in this audience. In contrast, other kinds of explanations, depicting causal patterns in, say, the role of genetic drift or phenotypic plasticity, are not good explanations for this audience. Given this specific deployment of the aim of understanding, the idealization that a population is infinite in size can be epistemically acceptable and help generate understanding. To be more specific, this idealization is epistemically acceptable so long as it is accurate enough to help represent the actual causal role of selection. Drift might still be a significant causal influence on the evolutionary outcome; it simply must not obliterate the pattern of selection's causal role. In contrast, the same idealization would not be epistemically acceptable if researchers instead wanted to understand the role of drift in the focal evolutionary outcome or the interplay of drift and selection. Thus, whatever values contribute to the prominence of studies of natural selection in population biology and guide individual researchers to take up those studies have tangible epistemic effects. Those values help determine both what causal patterns yield understanding and what posits qualify as epistemically acceptable.

The particular ways in which a research program and the values that shape it influence the nature of our understanding can be difficult to discern. For this reason, scientists interested in the same or similar phenomena sometimes construe their different approaches as stemming from opposed commitments to different states of affairs, when instead the difference can be traced back to their different interests. I briefly discussed this in Chapter 4, as part of my treatment of how science's different aims and the different deployments of those aims are pursued separately. There I referenced the disagreement in population biology regarding the significance of natural selection versus other evolutionary influences. A famous paper by Gould and Lewontin (1979) ushered in an era of polarization in evolutionary biology between so-called adaptationists and their critics. Gould and Lewontin criticized biologists who approach evolutionary biology by identifying some trait in a population and investigating how natural selection might have led to that trait. They called this approach "adaptationism" and pointed out that it neglects a number of other important influences on evolution. In its present form, this criticism primarily targets the use of optimality models and game theory to treat evolutionary phenomena, as these both represent the role of selection to the exclusion of other influences. Optimality models and evolutionary game theory have equally committed proponents and opponents among evolutionary biologists. Brown (2001) defends these approaches, together with a commitment to the importance of natural selection, as his "worldview." Along the same lines, Mitchell and Valone (1990) represent the debate as a choice between embracing either the assumptions of evolutionary game theory or those of quantitative genetics.[2]

My emphasis on how specific research programs influence the understanding they produce suggests a different construal of this divide in population biology and of similar disagreements. One need not hold that optimality models and evolutionary game theory best capture all important evolutionary influences to defend their scientific value. To the contrary, I have argued that these models are best seen as representing the causal role of natural selection in particular, at the expense of their accuracy of other causal influences or, indeed, of specific phenomena.[3] Biologists know too much about other influences on evolution to truly subscribe to the notion that natural selection is the only evolutionary influence or the only causally significant evolutionary influence. Some biologists do claim that natural selection is the only *important* influence, but this doesn't hold up as a claim about causal significance. It must instead be interpreted as a preference for tracking that causal pattern instead of others. And this is simply a way of specifying one's research interests. Evolutionary game theory and quantitative genetics, to name just one alternative, each may accurately represent different causal patterns, sometimes embodied by the same evolutionary outcomes, while misrepresenting other causal contributors with the use of idealizations. Each furnishes the proper explanations for its audience and, accordingly, full-fledged understanding of relevant phenomena.[4] These research programs are not in opposition to each other; they merely respond to different interests.

This reveals the great epistemic significance that values can have, simply in virtue of their apparently mundane role of shaping the research agenda. Researchers' different concerns and values can help lead them toward or away from a focus on natural selection's role in evolution. Researchers' shared concerns and values, including unexamined values, might result in the field of population biology disproportionately emphasizing the significance of selection and neglecting other evolutionary factors. This was the situation Gould and Lewontin diagnosed in late twentieth-century biology and dubbed adaptationism. In both cases, neither the avenue by which values exert their influence nor the nature of the values themselves is problematic. By setting the research agenda, values simply influence the type of understanding that is sought, what limited humans individually or collectively want to understand about the phenomenon under investigation. The only mistake is to neglect to appreciate the impact human concerns have on the nature of our scientific understanding. Recognizing this human-centric element of scientific epistemology enables the proper diagnosis of persistent deep divides among researchers, like those who focus on natural selection versus genetics and those who focus on environmental influences on aggression versus neurological influences. And it also enables the awareness that a field's limited focus may not reflect the state of the world so much as researchers' myopia. Such awareness supports the increasing appreciation of both scientists and philosophers of science for the importance of diversity of all kinds among scientists.

So far I have focused on social values' influence on the choice of research program, but features of scientists and their target audiences are epistemically influential in other ways as well. Even within the confines of a specific research program, what a scientist brings to her work can influence which causal influences are salient, what patterns are recognized in the phenomena, and thus the variety of understanding produced. Consider as an example Hrdy's (1986) analysis of how a shift in the gender ratio of primatologists led to significantly different types of observations about the behavior of male and female primates. This was discussed in Chapter 1. Different researchers, with different backgrounds and focuses, may notice different features of a phenomenon. In this case, a greater proportion of women primatologists generated a new focus on features of female primates that had been wholly neglected and that did not conform to the dominant theory. Hrdy says, of her own research trajectory, "First came an unconscious process of identification with the problems a female langur confronts followed by the formulation of conscious questions about how a female copes with them" (149). Here is another example not far removed from Hrdy's discussion of primatology. Roughgarden (2009) is explicit that the hypotheses and modeling approaches she calls "social selection" (in contrast to sexual selection theory) are inspired by her commitment to the idea that kindness and cooperation are "basic to biological nature" (1).

Features of the intended audience for some research—such as their background knowledge, their interests, and their position to act—can also shape the variety of understanding scientists aim to provide. Longino (2013) shows that research into the genetic influences on human behavior receives a disproportionate amount of popular attention, compared to research into environmental influences. Any number of features of the popular audience might contribute to this. There may well be a widely shared preference for simple, reductive explanations or for accounts of behavior that emphasize inheritance. Perhaps genetic influences on behavior are more surprising and accordingly more interesting than environmental influences. In my view, this situation is not in itself problematic. What is problematic is attributing the focus on genes to how the world really is, to the relative causal importance of genes over environment, rather than to simply what human observers have most often sought to understand. It is also problematic if there is a mismatch between the interests of the broad audience for science and the interests of those who control the research agenda, including, perhaps most significantly, those who control research funding.

How the values and perspectives of scientists and their audiences guide research often has been ignored by philosophers because of its apparent irrelevance to epistemic concerns. But there is now a well-developed literature, largely in feminist philosophy of science, about how features of the practitioners of science influence the content of scientific knowledge. The view developed here sets out one way in which that influence is exerted. Which research program is pursued, by whom, and for what audience all shape the character of the understanding that is sought, and so all have epistemic significance. These help determine which causal patterns are focal and which idealizations are employed. Social values may not directly influence whether a claim is true, but they do influence whether a causal pattern generates understanding, as well as whether a posit is epistemically acceptable and thus can contribute to legitimate scientific understanding. This is, then, one of the ways in which the seemingly humble influences of our social values on the research agenda are embedded in our scientific products.

As I indicated at the outset of this section, science's diverse aims and the different products that serve those aims extend the influence of social values on science beyond the epistemic influence on understanding I have characterized so far. In Chapter 4, I motivated the idea that science's aims and the specific deployments of those aims are best served by different scientific products. I opposed this idea to the view, often left implicit, that the same true or approximately true representations are used to accurately predict, to explain phenomena, to guide policy decisions, and so on. If this were the case, the influence of social values on scientific products would be constrained by the inability of values to influence what is and is not true. What scientists aimed to predict or evidential standards for policymaking, for instance, could be influenced by our values, but what representations best satisfy those aims could not be. Embracing instead the idea that success with one aim tends to inhibit successful pursuit of other aims eliminates this constraint. The most predictively accurate representations are not very explanatory, our best explanations tend to give somewhat inaccurate predictions, and so on. Values influence which aim and which specific deployment of that aim is pursued. This, in turn, influences what characteristics the resulting scientific product should have.

Here is a brief example of how values might influence scientific products via the precise deployment of a research aim (distinct from the epistemic aim of understanding, which I already addressed above). The political commitments of researchers or funding agencies might influence whether predictive research on climate change places greater emphasis on the riskiest scenarios, the least risky scenarios, or weights scenarios simply by their likelihood. These are distinct predictive aims that will generate different results. Indeed, Winsberg (2010) argues that, at least for climate science, the choice of a specific prediction task is always influenced by nonepistemic considerations, including values. Social values cannot influence whether a prediction is accurate, just as they cannot influence whether a causal pattern obtains. But by influencing the exact nature of the predictive aim, values can shape the nature of predictions analogously to how they shape the nature of our understanding. This idea of how values legitimately shape scientific products through their influence on highly specific research aims is similar to the conclusion reached by Elliott and McKaughan (2014) that nonepistemic considerations can override epistemic considerations in virtue of scientists' different goals, when coupled with trade-offs among the desirable features scientific representations might possess. It is also reminiscent of Longino's (2001) suggestion that even the epistemic value of empirical adequacy, which might seem basic to all empirical pursuits, is actually negotiable, for some accuracy can be sacrificed for other desirable features, like simplicity or generality.

The values that influence science by determining the nature of understanding sought, or by otherwise shaping the research agenda, can vary across scientists and research programs. But even with such variation, the range of values that shape the nature of our scientific understanding will always be partial and—I expect—particularly human, perhaps even particular to moments of human social and intellectual history. This is inescapable, and it need not be epistemically or ethically problematic. This is what makes our science truly ours. But this does open up another dimension of scientific progress. The range of values that shape our science should be broad enough to reflect the concerns and values of the full human audience for science. The practitioners of science should thus be multiply diverse, and the diverse audiences for science should be recognized and given voice. This is a very condensed and general statement on a matter worthy of extended consideration, a matter that has been given extended consideration by other philosophers. This is the question of what values or range of values should influence our scientific aims and the scientific products that serve those aims. One influential view of this matter is developed by Kitcher (2001, 2011), whose account of "well-ordered science" is designed to govern what kinds of values should influence the research projects pursued and the standards of acceptance for scientific findings.

My purpose in this section has been to show how some of the views developed in this book elevate simple ways in which values influence the research agenda to great significance, including epistemic significance. Two individuals cannot know contradictory things about the same phenomenon, defined in the same way. This is so even if those individuals have different backgrounds, values, interests, and so on. But all of those differences among the individual practitioners and consumers of science *are* relevant to what phenomena are investigated, how those phenomena are construed, which causal patterns are focal, what connections are drawn to other phenomena, and other elements of the research agenda. All of these differences in turn affect the variety of understanding that is sought or, more generally, the scientific aim and specific deployment of that aim pursued. This provides a means for our characteristics and values to legitimately influence the nature of our scientific understanding in a way that they cannot influence truth. Note that this is not supposed to be a complete account of how values influence science but merely an accounting of the implications the account of the aims of science developed in this book have for that issue.

In Chapter 4, I defended the idea that taking understanding to be the epistemic aim of science is not problematically subjective. Let me briefly revisit that issue now, in light of what I have suggested here regarding social values' epistemic influence. The avenues I have identified for the influence of values correspond to ways in which epistemic success and the achievement of science's other aims are properly subjective. What best facilitates human understanding and other aims humans set for science depends in part on the features of those humans. And yet, just as our values cannot determine whether a hypothesis is true, there are regards in which social concerns cannot legitimately influence the content of our understanding or other scientific products. Once the focal phenomenon and research agenda are set, we cannot choose whether a given idealization facilitates understanding or whether some representation can generate understanding. Whether a causal pattern is embodied and whether the requirement of epistemic acceptability is met cannot be determined directly by our cares and concerns. Epistemic acceptability provides only a loose and variable tether to reality, but it is a tether nonetheless. The same goes for the pursuit of other aims. Whether a prediction is accurate is immune to our desire for it to be so. And yet scientific success is determined jointly by a combination of ingredients. These include not only the features of phenomena under investigation and how we represent them but also the precise nature of the research program pursued, as well as the human psychological characteristics and concerns that motivate that research.

7.2 How Science Doesn't Inform Metaphysics

The rise of naturalistic philosophy—that is, philosophy that accords some or even total authority to empirical science—has been accompanied by questions about the relationship between science and metaphysics in particular. Attempts to naturalize metaphysics aim to demonstrate how our best scientific findings should inform our metaphysical commitments. In this section, I suggest restrictions on science's metaphysical implications that derive from the account of science's aims and methods developed in this book. A handful of specific limitations on science's metaphysical significance were already indicated in earlier chapters. In Chapter 2, I argued that an account of how the concept of causation is employed in science may well not give direct insight into the metaphysics of causation. In Chapter 6, I defended a similar conclusion for levels of organization: that the targets of scientific investigations and science's organization into fields of study do not correspond to any metaphysically significant levels of organization. And, in Chapter 4, I argued that the epistemic aim of science is not best conceived as truth (or increasing accuracy, or similar). The purpose of this section is to explicitly consider the question of the metaphysical significance of our best scientific findings in light of these limitations.

Let me be clear that I do not pretend to settle the issue of the proper relationship between science and metaphysics here. Many philosophers have drawn a variety of connections between science and metaphysics, and much effort has gone into articulating and analyzing versions of naturalized metaphysics. I engage directly with only a few examples of drawing metaphysical implications from science. My remarks may be irrelevant to some other approaches to a naturalized metaphysics, and they may be fully consistent with yet other approaches. As in the previous section, my aim here is only to trace out the implications of positions I have defended in this book for a further philosophical question about science. In particular, my account of idealization and the aims of science motivates some forms of caution in examining science for metaphysical import.

I have argued that idealizations are rampant and unchecked in science. This means that not only are idealizations widespread, but also little is done to control or limit their influence on our representations. Significant causes can be idealized simply because they are unimportant to researchers' immediate interests. This view of the centrality of idealizations in science in turn motivates my account of the aims of science. There are a variety of scientific aims, these aims are best pursued separately, and the epistemic aim of science is not truth but understanding. These views all curtail science's direct metaphysical import. Representations with unchecked idealizations and the understanding they generate are not well suited to ground metaphysical conclusions. I have argued that scientific understanding is yoked to particularities of our epistemic position and of human psychology. These are two limitations that it seems any successful metaphysics must escape. That science has a variety of aims best pursued separately is similarly limiting. This undermines the idea that scientific products are converging—or will converge at some future point in time—and will then provide a unified account of the world. A diverse array of tools for piecemeal prediction, explanation, policy guidance, and so on holds little promise of direct metaphysical import.

There is also a broader version of this argument against science's direct metaphysical import. As we have seen, science is a human endeavor. It is a tool, or really several tools, designed by humans to further our cognitive ends and our ability to exert influence on our world. Furthermore, in the previous section, I detailed some ways in which human features and values enter into science. If I am right that science's status as a human enterprise and even the specific values of its practitioners indelibly shape its character, then scientific findings are simply not objective in the right way to directly inform metaphysical conclusions. Surely any metaphysics aspires to escape the particularities of human concerns, cognitive abilities, and shared or individually held values. This, coupled with the considerations introduced in the previous section, suggests that our metaphysics cannot be read off the products of science.

It is thus not incidental that the account of the aims of science developed in this book involves little in the way of metaphysical commitments. I urge that we embrace facts about manipulability as epistemically basic to any other forms of causation, not as the metaphysical basis of causation. I argue against the scientific significance of composition, realization, and universal levels of organization, not the metaphysical absence of them. I have said that causal patterns are real and that these are the objects of our scientific knowledge. But by the reality of causal patterns, I simply mean that there are objective regularities in manipulability relations. This is intended to contrast with the idea that those patterns are imaginary or subjective. I can't see how it could matter to the positions developed in this book if one is a nominalist or a realist about causal patterns or even if one resists calling these patterns causal, preferring instead the nonpartisan term "manipulability patterns." In my view, we can have knowledge about causal patterns, but this is a kind of secondary good, valuable as the route to scientific understanding of phenomena. Even the scientific realism my view enables must be construed in terms of the epistemic success of science, which for me consists in nonfactive and interest-relative understanding, emphatically not that the posits of our best science are uniformly—or even largely—believed.

The idea that our scientific products are not well positioned to provide metaphysical insight will become clearer by considering a few ways in which science has been used to motivate metaphysical conclusions and evaluating how these fare in light of the views developed in this book. Dupré (1993) suggests that, even though metaphysical presuppositions ground the enterprise of science, scientific findings can and should be used to corroborate or undermine those assumptions. Metaphysical positions on the chopping block include essentialism about natural kinds, reductionism regarding entities and scientific theories, and determinism. In the end, Dupré concludes that his conception of scientific disunity implies the failure of all three metaphysical theses. In this book, I defend an account of science that is friendly and in some regards similar to Dupré's view of science. But the implications I see for metaphysics are different in kind. I think one can consistently recognize the features of science Dupré attends to and yet resist his metaphysical conclusions.

Consider natural kinds. Positions developed in this book accord with Dupré's (1993) conclusion that science involves a multiplicity of cross-cutting categorizations. Dupré embraces the idea that these scientific categorizations are truly natural kinds, and he is a realist about them. He calls this "promiscuous realism" about natural kinds. But note that whether we should be realists about natural kinds and, indeed, whether scientific categorizations capture natural kinds cannot be gleaned from science itself. Dupré's conclusion that science undermines essentialism about natural kinds would be better put as a conclusion about scientific kinds. One who endorses realism and essentialism about natural kinds could grant this as a point about categorization schemes in science but look elsewhere for natural kinds or take them to be unknowable. Alternatively, one might agree with Dupré about scientific categorization and take this as reason to reject the reality and naturalness of kinds all together. Dupré's conclusions about science are compatible with any of these metaphysical positions.

I also agree with Dupré that scientific practice undermines reductionism about scientific theories and other representations (see Chapter 6). This is a thesis about the enterprise of science. But Dupré also addresses the metaphysical question of the reduction of entities, and this cannot be settled simply from a consideration of science. A position might be informed by our best scientific findings, such as science's general failure to uncover realization relationships (see Potochnik 2010b). But one might instead grant this while still maintaining metaphysical reductionism about entities. One could simply attribute the failure of science to hand us the realization relationships needed for metaphysical reduction to science's many competing goals. Perhaps when realization relationships are not identified, this is simply because that identification is not the immediate scientific aim and does not contribute to the immediate aim. The state of our science thus cannot arbitrate for us the question of metaphysical reductionism. There are more philosophical questions to ask, more philosophical arguments to develop.

Ladyman and Ross (2007) urge that science be taken as the basis for metaphysics, and they emphasize science's ability to get at the "objective character of the world." Their criticism of classical metaphysics disengaged from science is, in part, based on the idea that when common sense and science conflict, commonsense ideas must be jettisoned in favor of science. On their view, fundamental physics is also privileged over all other fields of science. Ladyman and Ross endorse an exceptionless principle to this effect. As quoted in Chapter 6, they say that a hypothesis conflicting with findings in fundamental physics is sufficient reason to reject that hypothesis. As a result, they hold that quantum physics has demonstrated that there are no things, only structures.

First let us consider whether scientific findings always warrant jettisoning commonsense ideas. Science's differences from everyday reasoning are essential to its power and success, but I have suggested that those differences ill-suit it for doing direct metaphysical work. To consider an example relevant to Ladyman and Ross's project, quantum physics is not reconciled with the theories of relativity, and I do not know of a reason to expect it to be in the future. Each of these scientific achievements provides an understanding of a certain domain—one domain including certain phenomena at very large scales, the other certain phenomena at infinitesimal scales.[5] Multiplicity and variety tend to persist among our scientific approaches and findings. This was illustrated in Chapter 3 with the case studies of various approaches to studying human aggression, the variety of behavioral ecology approaches to accounting for cooperative behavior, and even the need to appeal to classical orbits to explain some quantum phenomena. Assumptions and even direct claims that figure into different scientific achievements are often not reconcilable with one another. There are thus no grounds for concluding that those assumptions and claims always should be reconcilable with extra-scientific beliefs. The ways in which science's practices and products are human-centric prevent them from being wholly objective in the strong sense needed to ground metaphysical conclusions.

Melnyk (2013) is critical of Ladyman and Ross's idea that science provides access to objective reality. Nonetheless, he agrees that when the results of scientific methods and everyday methods conflict, we should always prefer the former. But this is also wrong. Science extends the reach of human understanding considerably, but it cannot escape its basis in our collective experiences. Those experiences must ultimately ground our concepts and our scientific understanding. Of course, this is not to suggest that quantum mechanics should be rejected because it belies what common sense would lead us to expect. But it is overly hasty to conclude on the basis of quantum physics alone that there are no objects (at any scale, on any construal). That idea is not an element of quantum physics but an extension of it to metaphysical conclusions about middle-sized objects. It is familiarity with those middle-sized objects and the scientific and nonscientific generalizations that can be made about them that serve as our epistemic point of departure for all of science. Extending scientific results in certain ways without substantial additional philosophical argument amounts to an overextension; I suggest this is one of those ways.

A brief departure from the main topic here will, I think, prove helpful. The relationship I have motivated between science and commonsense reasoning is reminiscent of an idea central to Otto Neurath's conception of science and, in particular, the role of so-called protocol sentences. In Neurath's view, protocol sentences—the ultimate epistemic basis for all of science—are third-person reports of everyday observations, couched in ordinary language. That language is ambiguous and imprecise; Neurath called its terms *Ballungen*, which has been variously translated as "congestions," "conglomerations," and "clusters" (see especially Cartwright et al. 1996). The ambiguity and imprecision of these terms serves a purpose: they stabilize language use across individuals, cultures, and time periods. In contrast, scientific terms tend to be more precise, but they are also theory driven. Neurath points out that "the terms of science must adapt themselves much more to the new theories than a cluster [*Ballung*]" (1936a, reprinted in Neurath 1983, 149). And so, according to him,

> Our whole life consists in two opposite movements: in the one we tend to acquire always new concepts and to modify those that tradition has left us; but in the other we are obliged to take the traditional statements as the basis for our departure. (Neurath 1936a, 150)

This idea is akin to and indeed related to the epistemic relationship I have suggested between common sense and science. The tools of science have much extended the reach of human understanding, and they have rightly led us to reject many commonsense beliefs that at one time appeared unassailable. But some of our commonsense beliefs get the whole endeavor up and running, and even the best results of our scientific enterprise will always be more tentative.

I also take issue with another position Ladyman and Ross use to support their metaphysical conclusions. In my view, physics should not be accorded a special status relative to the other sciences. This is a claim about the relationship among the fields of science, and so we should be able to adjudicate on it on the basis of observations of science. But what science shows us does not favor Ladyman and Ross's conclusion. I have already argued in Chapter 6 that the field of microphysics is not epistemically privileged in virtue of its focus on the very small. Instead, the epistemic position of our best fundamental physics is similar to or perhaps worse than our best findings in other fields. These investigations face a number of epistemic difficulties that other fields may not, such as the unavailability of direct observation and the tremendous amount of equipment generally required for experimentation. There is simply no reason for our best physical theories to override findings from other fields.

Consider that William Thomson, later Lord Kelvin, and Darwin famously disagreed regarding the age of the Earth. In the first edition of the *Origin of Species* (1859), Darwin estimated from geological evidence that the Earth was of an age sufficient for gradual evolution by natural selection. However, based on thermodynamics, Kelvin estimated that the Sun was 30 million years old. This suggested that the Earth had actually not been habitable for a sufficient period of time for such evolutionary change. Darwin was sufficiently convinced by this that he removed discussion of timescales from the later editions of the *Origin*. But it was Kelvin who was wrong: he did not appreciate the role of fusion in the Sun's production of energy. The well-founded, carefully applied physics turned out to be wrong, and the geology and speculative new biology turned out to be right. Granted, certain claims in physics are more epistemically secure than certain claims in other sciences. But the reverse is also true. If I am right that physics is not more trustworthy than other fields of investigation, then this undermines those projects, like Ladyman and Ross's, that would base their metaphysical conclusions entirely on the products of fundamental physics. In biology, geology, and many other properly scientific fields, references to particular objects abound.

Ultimately, my point in this section is a modest one. The position I advocate is simply that caution, as well as significant philosophical argumentation, is warranted when drawing metaphysical conclusions from the practices or products of science. One cannot in general simply read metaphysical implications directly off scientific findings. This is due to the nature of science—its human-centric design and its limited connections to truth attenuate its metaphysical significance. An extension of scientific findings to inspire some metaphysical position does not necessarily trump commonsense judgments; this depends on the details and requires additional consideration and argumentation. And finally, physics is not special among the sciences in method or epistemic security. None of these ideas bars a naturalized metaphysics or a "scientific" metaphysics. That I advocate a wedge between scientific findings and metaphysical implications does not mean I think analytic metaphysics should ignore the scientific enterprise and continue business as usual. Metaphysical projects almost certainly benefit from aiming for consistency with and guidance by our best science (as well as everyday reasoning), and many metaphysicians do just that. But metaphysics cannot be read directly off our science, not even our fundamental physics.

7.3 Scientific Progress

Some of the views I have put forward in this book may seem as if they threaten to undermine the legitimacy of the scientific enterprise. I have suggested that there is rampant and unchecked idealization in science, that scientific understanding is facilitated by departures from the truth, that there is a direct role for social values in shaping scientific products and in the epistemology of science. These and related views seem to combine to form a rather radically human-dependent view of science. One might reasonably be concerned that science, on the view I have urged, is not well positioned to accomplish the epistemic aims and other functions for which humans designed it. And yet, at the conclusion of Chapter 1, I promised that the view of idealization and the aims of science articulated in this book does not undermine the success of science but rather brings its successes more fully into focus. In this final section of the book, I show how this is so. I first analyze the modest scientific pluralism and its limits that my account supports. I then suggest a strong interpretation of scientific progress and thus of the cumulative epistemic success of science. I close by outlining some prescriptions for scientific practice that follow from the interpretation of science's aims set out in this book.

Notice, first, that the positions I have advocated generate an enduring pluralism regarding scientific aims and products. I have suggested that idealization is widespread but that those idealizations are useful for a variety of different reasons and to a variety of different types of scientific projects. As a result, the various aims of science are in tension with one another, in the sense that they motivate different idealized products. Even the specific aim of understanding—the primary epistemic aim of science, but still only one aim among many—can motivate multiple different idealized representations of the same phenomenon. This is because which representation facilitates understanding depends on which features of a phenomenon are focal and, accordingly, which causal pattern is explanatory. I have called this variation different deployments of the aim of understanding. Earlier in this chapter, I discussed how our individual and shared social values are reflected in this range of scientific aims and products. And in Chapter 6, I challenged the basis for ordering our scientific products according to their "level." All of these are reasons to expect an astounding variety of approaches and products in science, that this variety will continue unabated, and that the various scientific products are by and large not reconcilable or combinable in any fruitful or enlightening way.

On the other hand, my version of scientific pluralism is a modest one. Perhaps surprisingly, some of the views advocated in this book preclude or limit implications of pluralism that might otherwise seem to threaten the epistemic security of science. That there are a variety of scientific findings, some of which appear to be inconsistent, may seem to undermine the legitimacy of those findings, if not the scientific enterprise as a whole. Otherwise, such inconsistent findings may seem to require abandoning belief in a single, unified reality; the principle of noncontradiction; or some other radical maneuver. But none of this follows from my account of the aims of science. I have argued that science does not aim for truth; a continuing pluralism of scientific results thus does not undermine the epistemic legitimacy of science or of any of its individual results. Apparently inconsistent scientific findings, on closer inspection, often are the result of different types of aims or different deployments of an aim. Nor does continuing pluralism in science imply the absence of a single reality or even a single well-ordered reality. As I have repeatedly emphasized throughout this book, the mixture of causal complexity and human cognitive needs leads to the variety observed in our scientific products. This mixture contributes to science's limited direct metaphysical import, and, in particular, it undermines any conclusions about metaphysical disunity or disorder from scientific pluralism.

This view contrasts with some prominent views of scientific pluralism and disunity, including notably Dupré (1993) and Cartwright (1999). Both have different ideas about the metaphysical implications of scientific pluralism. I discussed Dupré's ideas about the metaphysical implications of scientific findings in the previous section. Cartwright (1999) also develops an account of the disunity of science. She rejects the idea that science is working toward a unified fabric of representation and maintains instead that there is a plethora of scientific laws, with no systematic relationship among them. I agree with her on this much. Cartwright also suggests that it is plausible that nature turns out to be "constrained by some specific laws and by a handful of general principles, but it is not determined in detail, even statistically" (49). In this claim, she is not *asserting* metaphysical disunity on the basis of scientific disunity. But, by finding it necessary to defend the possibility of metaphysical disunity, it seems Cartwright envisions a closer relationship between the state of our scientific laws and the state of the universe than I expect there is. In my view, our scientific products do tell us something about reality, but the relationship to actual phenomena is indirect, for it is mediated by causal patterns that are imperfectly embodied in a limited range of phenomena and only the particular patterns that we humans collectively or individually find enlightening. Positing scientific pluralism—one kind of scientific disunity—thus does not require positing or even allowing for metaphysical pluralism or disunity.

So, on my account of the aims of science, modest scientific pluralism does not indicate shortfalls of science or metaphysical disunity, since apparently inconsistent scientific findings often result from different aims and deployments of aims. However, this does not render scientific findings unassailable. Also in contrast with some versions of pluralism, the views developed in this book support the ability to assess scientific approaches and, more significantly, ground a nuanced form of assessment. The assessment and criticism of scientific methods and products are not curtailed by allowing for rampant and unchecked idealization or by the associated distancing from truth. It is the indexing of scientific success to a broad range of different aims, each of which might have a number of different deployments, that enables a plurality of scientific products. But this means that once a specific aim is identified, it is possible to evaluate the success of a given product in achieving that aim. In Chapter 4, I worked through how astrology falls short of its own representational aim. It is thus an empirical failure. How this empirical failure has been accommodated by astrology advocates is, in turn, a methodological failure.

Consider another example of the aim-specific evaluation of a scientific product, this one in the vicinity of successful science. I have discussed the use of the prisoner's dilemma to account for the evolution of cooperative behavior. I argued that this approach should be construed as aiming to provide an understanding of how natural selection can lead to cooperative behavior, and I gave an example of its application to food sharing among vampire bats. Given this aim, the prisoner's dilemma should not be criticized for failing to accurately represent the genetic influences on cooperative traits, and in fact these influences are typically wholly idealized in behavioral ecology. In contrast, the prisoner's dilemma *should* be assessed according to how well it represents the role of natural selection, and it may come up short. I have suggested elsewhere that reciprocal altruism, such as food sharing in vampire bats, may appear to be a prisoner's dilemma when it instead evolves by the direct selective advantage of individuals. In that case, there would be no individual cost, no selective disadvantage, to cooperating (Potochnik 2012). Taking into account specific aims facilitates nuanced assessments of scientific success. It can be difficult to discern what aim is applicable to some scientific finding or approach. But that difficulty does not necessitate or warrant abstaining from assessments. The recognition that scientific products serve specific aims instead provides a roadmap for how to conduct nuanced evaluations of those products and the approaches that generate them.

This seems to be a point of contrast with Longino's pluralism. Criticism and uptake of criticism are integral to Longino's contextual empiricism, but some features of her view may render critical assessment difficult. She holds that science is divided into numerous separate communities with differing epistemic values and that these communities generate incommensurable scientific approaches and findings (see especially Longino 1990, 2001). On its face, this incommensurability undermines critical assessment. This tension is signaled by Longino's (2013) discussion of approaches to researching human behavior, for she suggests at various points that if a scientist or philosopher criticizes some scientific approach, then the individual in question is a monist, that is, rejects scientific pluralism (e.g., 125, 138). The idea that criticizing an approach amounts to monism seems to stem from Longino's commitment to incommensurability of approaches and products across scientific communities. If approaches are truly incommensurable, then this may render external critique impossible. But this is, of course, in tension with Longino's requirement of critical interaction or at least with critical interaction across different scientific communities. Critical interaction among approaches seems to require direct comparison and communication, while incommensurability suggests the inability of one approach to speak to another.

From the perspective of the views developed in this book, Longino's emphasis on incommensurability should give way to the possibility of nuanced critique, including by outsiders. One may be a pluralist and yet still find fault with one or another scientific approach or product. Consider two examples from the investigations of human aggression addressed in Chapter 3. First, as Longino points out, behavioral genetics has been thoroughly criticized for a variety of methodological problems, as well as for conclusions that regularly outstrip their evidential basis. The approach of behavioral genetics may be methodologically flawed—even to the point of warranting its abandonment—without threatening scientific pluralism in general. From the perspective of the form of pluralism endorsed here, demonstrating this failure requires assessing the success of specific projects in the field of behavioral genetics in achieving their precise aim(s). Second, molecular genetics has been criticized for its lack of illuminating results. Recall from Chapters 1 and 3 the problem of missing heritability. If this difficulty proves intractable, it may warrant abandoning molecular genetics as an approach to investigating human behavior. But in this case, scientific pluralism has not been undermined. Instead, this failure would tell us something deep about the causal patterns of our world, namely, that there are precious few clear patterns in how particular molecular genes influence human behavioral characteristics. As these examples illustrate, some approach may prove methodologically problematic or empirically fruitless to the extent that it warrants abandonment without threatening pluralism regarding methods or findings. The sources I have suggested for a modest scientific pluralism allow a plurality of approaches and findings while enabling criticism of both.

My account of the aims of science thus blocks scientific pluralism from implying metaphysical disunity or disorderedness, and it facilitates the nuanced assessment of scientific approaches and products by taking into account their specific aims and the expected inaccuracies and limitations that result. Both of these implications might be viewed as curtailing the disunity of science—its metaphysical import and its epistemic import, respectively. A third limitation on pluralism is also opposed to some ideas about the disunity of science. This limitation is simply the idea that there is a range of productive generalizations to be made about the approaches and products of science. A theme of this book has been diversity in science. I have endorsed a wide array of different aims and a plurality of legitimate approaches and products, even for a single target phenomenon. There can be a variety of different representations of a phenomenon, each best for a different purpose, and there can be a variety of explanations of any given phenomenon. Relationships among scientific representations are not well ordered and undermine the idea of levels of organization. Yet, despite all this diversity, the purpose of this book has been to develop a number of generalizations about science, all stemming from the initial commitments to causal complexity and a scientific practice that reflects human concerns. Those generalizations significantly constrain the diversity of scientific practices and products.

Here are the primary generalizations about science set out in this book. First, I have motivated the idea that science is in the business of capitalizing on causal patterns. Representations of causal patterns serve as our scientific explanations and, thereby, produce scientific understanding. Scientific research sometimes capitalizes on causal patterns in other ways in the pursuit of other aims, like prediction or policy guidance. But causal patterns are key throughout—or so I have argued. This relates to science's status as a tool in service of human cognition and action. Causal patterns are at the heart of both types of purposes. Second, I have argued that there is rampant and unchecked idealization in science. Idealizations occur throughout the scientific enterprise, and they are not neatly accounted for by appeal to one or a few motivations. Idealizations serve different aims, and they are warranted for a variety of different reasons, temporary and permanent. Idealizations interfere with the accuracy of our scientific representations, resulting in the epistemic aim of understanding diverging from that of truth.

That science capitalizes on causal patterns and involves rampant and unchecked idealization are two generalizations about the whole of science. A third generalization is the unitary account of explanation I have outlined. On that account, there are three features possessed by any successful explanation: it represents a causal pattern, including its scope; it is responsive to the prevailing research interests; and it satisfies the requirement of explanatory adequacy. Although this is a unitary account, the indexing of explanations to research interests results in a variety of explanations for any given phenomenon. But this also is linked to another generalization about scientific explanation: the communicative sense of explanation is basic. Explanations are the vehicles of human understanding, and their nature is indelibly shaped by this. A fourth set of generalizations regards the failure of universal levels of organization in science and the expectation that projects from different fields and subfields will regularly benefit from evidential connections, in virtue of what I have called epistemic interdependence. This epistemic interdependence balances the independence of different explanations of a single phenomenon (Potochnik 2010a).

I have argued that these generalizations about science support a modest scientific pluralism and place reasonable limitations on its extent. The form of this pluralism also suggests a promising articulation of scientific progress and science's present success. This is due especially to the emphasis I have placed on different aims and deployments of aims, as well as the centrality of causal patterns, especially to the epistemic aim of understanding. At the close of Chapter 4, I suggested that taking understanding to be the epistemic aim of science in the way I urge enables a cheery interpretation of scientific progress. Superseded scientific theories were by and large not true, but many of them did capture causal patterns and engender understanding. From this perspective, scientific progress is in some ways cumulative. We are ever amassing more understanding and success in other aims as well, even as what, on a more traditional view of science's aims, is posited as true is continually in revision. One way to formulate this point is that pluralism regarding successful scientific products extends over the history of science as well as in any given moment of science.

We can thus be confident of our present science's success, epistemic and otherwise. Even if our current models and theories are later abandoned by a future science, that is unlikely to undermine their current contributions to human understanding, to our store of knowledge about causal patterns.

Science doesn't get us truth—but it's not designed to. Instead, it delivers goods much more useful and enlightening for us than simple truths about our complex world could ever be. This account of idealization and the aims of science does not undermine the value or success of science. Rather, it supports a stronger articulation of science's success than other, more traditional accounts can hope to.

From the beginning of this book, I have emphasized my commitment to account for today's science, as actually practiced. But I have also suggested that this commitment is compatible with critically assessing science's approaches and products. Just above I discussed how my modest scientific pluralism enables criticism of approaches on both methodological and empirical grounds. It's also possible to critically access those approaches in the aggregate—to ask not just how science is in fact practiced but how it might be better practiced. My account of science's aims inspires some normative philosophical conclusions about scientific methodology. These follow closely on the heels of points I have already made. I thus conclude this book by drawing out some prescriptions for scientific practice based on the analysis I have conducted. A summary of these normative conclusions regarding scientific practice is provided in Table 7.1.

Let's begin with implications of the view that all scientific products are heavily idealized and, accordingly, only successfully contribute to specific, limited aims. For this reason, scientists, as well as the public audience of science, should anticipate a continuing diversity of scientific results and

TABLE 7.1. A summary of prescriptions for scientific practice based on the philosophical analysis I have conducted.

Philosophical conclusions	Prescriptions for scientific practice
1. Rampant idealizations, diverse aims: all scientific products are partial and idealized; different products satisfy different aims	Modest scientific pluralism: continued diversity of scientific products; irreconcilable but evaluable
2. Diverse aims shaped by values: idealized representations depend on aim; aims are in service of human goals	Entrenchment of social values: recognize how social values influence aims; pursue aims that reflect endorsed values
3. Failure of levels of organization: properties and entities are not universally stratified; hierarchy concepts do not cohere	Epistemic interdependence: jettison appeals to levels; seek out evidential interconnections
4. Our science reflects human concerns: science is structured to facilitate human action and cognition	Science for all humans: pursue research that furthers human action and understanding, wherever that leads

approaches, even when those results and approaches all deal with a single phenomenon. A modest scientific pluralism will forestall mistaken opposition among approaches that simply employ different idealizations in pursuit of different goals. An example of this mistaken opposition, discussed above, is provided by those who believe the success of evolutionary game theory entails the uselessness of quantitative genetics, as well as the opposed belief that the representational limitations of evolutionary game theory entail game theory's uselessness. And yet, recalling the earlier discussion of limits on pluralism, the scientific pluralism that results from heavily idealized, partially successful scientific products also inspires a form of interconnection. Each approach provides insight that can be used in the assessment of others, provided that the specific aims of different approaches are taken into account. The pluralistic products of science are, then, not incommensurable in my view. Multiple treatments of a single phenomenon simply are differently oriented to that phenomenon, in virtue of the different specific aims they serve.

Consider, second, the implications for scientific practice of the combination of two ideas I have defended: that which idealized representation is appropriate depends on the specific aim and that these specific aims are set in part by human goals and thus human values. I suggested earlier in this chapter that this elevates apparently mundane ways in which social values influence scientific research projects to epistemic significance, resulting in the entrenchment of values' influence on the actual products of science. From a scientist's perspective, this view has at least two implications. First, that it is important to recognize how social values influence the specific aims of one's research and, second, to choose one's specific research aims in a way that reflects the values one chooses to endorse. A perfect example of this is provided by Joan Roughgarden's work, which has been discussed in a few places elsewhere in this book. Roughgarden (2009) analyzes how elements of sexual selection theory reflect expectations of opposition among individuals' self-interest, discord between the sexes, and dishonesty. She then produces a series of alternative hypotheses to the broadly accepted views of sexual selection theory, explicitly designed to reflect the expectation of mutual benefit, cooperation, and successful communication. In her first sentence, Roughgarden declares that "this book is about whether selfishness and individuality, rather than kindness and cooperation, are basic to biological nature" (1).[6] I suggested above that part of what differentiates Roughgarden's hypotheses from sexual selection theory is a focus on different causal patterns. This difference in focus derives from different specific aims—and Roughgarden's aim is transparently influenced by values she explicitly endorses.

Third, my argument against universal levels of organization also has direct implications for scientific practice. I have suggested that philosophers should abandon talk of levels of organization entirely, for even when a specific levels concept is carefully developed and employed properly, the weight of the concept's history is likely to lead to overextension and unprincipled usage. I think the same is true for scientists. Appeals to levels in science should be replaced with more careful and limited reference to scale relationships, composition, mechanical organization, and so on. Appeals to levels have been used to justify problematic assumptions in, at least, physics (Rueger and McGivern 2010), ecology (Potochnik and McGill 2012), and individual-based modeling in the social sciences (Epstein 2012). If there are no universal levels of organization, then the fields and subfields of science are not ordered according to the levels of reality they address. I have suggested that, instead, these fields are related by coordination. Causal complexity results in phenomena with relevance for multiple fields of science, and even though phenomena are targeted differently by different fields—indeed, because of this—these approaches regularly benefit from collaboration. Research programs benefit from taking into account investigations of similar phenomena, even with different aims, in part because different investigations employ different idealizations. Scientists should then seek out such epistemic interconnections, even as they embrace pluralism.

Fourth, and finally, my focus on how science is shaped by specific human characteristics and concerns can be used to motivate the explicit adoption of this as a standard governing scientific success. Our science is, ultimately, a tool to facilitate human action in and understanding of our causally complex world. The aim of action motivates departures from universal generalizations, the aim of understanding motivates departures from truth, and both limit the direct metaphysical insight our science provides. Keeping these limitations in mind can contribute to a restructuring of science's research priorities. For example, this undermines any prejudice in favor of investigating the action of the small instead of the large, or of components instead of wholes. Scientists should pursue research that furthers human action and human understanding. This suggests a focus on useful and enlightening patterns, wherever they are found, and on factors upon which humans can intervene. This is especially so when the phenomena are of human ethical concern, such as global climate change, human physiological and psychological health, and cooperative action.

Acknowledgments

First and foremost, I want to thank my amazing colleagues at the University of Cincinnati for their help with this book, especially Zvi Biener, Vanessa Carbonell, Anthony Chemero, Valerie Hardcastle, John McEvoy, Thomas Polger, Robert Richardson, Robert Skipper, and George Uetz in biology. Guilherme Sanches de Oliveira provided research assistance and help with indexing. I also received tremendously helpful feedback from the graduate students and faculty who participated in my seminar on idealization in the autumn of 2014, including some of those named above, as well as Frank Faries, Maurice Lamb, Vicente Raja Galian, Walter Stepanenko, and Richard Stephenson. I am fortunate to be surrounded by such talented and generous colleagues and students; this project was significantly shaped by their input and criticism.

I also availed myself of the generosity of colleagues and mentors at other institutions for their guidance and feedback on this project, including especially Adrian Currie, Elihu Gerson, Eric Hochstein, Arnon Levy, Helen Longino, Elliott Sober, Michael Strevens, Alex von Stein, Michael Weisberg, and Cory Wright. I'm fairly certain Adrian Currie provided the most detailed, constructive feedback any book manuscript has ever received. And Michael Strevens's feedback vastly improved both Chapters 2 and 4.

The ideas put forward in this book were also significantly influenced by questions and comments from audiences of talks at Case Western Reserve University; Indiana University; Saint Louis University; the 2015 meeting of the International Society for the History, Philosophy, and Social Studies of Biology; the Missouri Philosophy of Science Workshop in 2013; Models and Simulations 6 at the University of Notre Dame; the 2014 Pacific Division Meeting of the American Philosophical Association; the 2014 Biennial Meeting of the

Philosophy of Science Association; and a 2014 workshop on modeling and scientific explanation at the Van Leer Institute in Jerusalem.

I also owe thanks to the Charles Phelps Taft Research Center at the University of Cincinnati. I drafted this manuscript during a Taft Center Fellowship in 2013–2014, and other Taft awards supported much of the research that has found its way into this book. Thanks also to the University of Chicago Press, as well as Karen Darling and Evan White for their guidance and help. The book was also improved by comments from three anonymous reviewers for this press and two for another.

Finally, I greatly appreciate the love and support of my family: my parents Debbie and Tony, who have endless patience for my esoteric philosophical ideas but only wish those ideas could help just a little bit with climate change; my husband Andrew, who is still worried about the title of Chapter 4; and my daughters, Mabel and Amelia. Mabel has asked me to give copies of this book to her friends but has also said she won't read it because it doesn't have enough pictures.

This book contains revised text from some of my previously published articles. A revised version of "Scientific Explanation: Putting Communication First," published in *Philosophy of Science*, contributes to sections 5.1 and 5.2.2. "Causal Patterns and Adequate Explanations," published in *Philosophical Studies*, has been significantly revised and forms the basis of section 5.2. A significant portion of "The Limitations of Hierarchical Organizations," coauthored with Brian McGill and published in *Philosophy of Science*, has been revised and included in sections 6.1 and 6.2. Some revised passages from "A Neurathian Conception of the Unity of Science," published in *Erkenntnis*, appear in section 6.3. A revised version of much of "The Diverse Aims of Science," published in *Studies in the History and Philosophy of Science*, appears in parts of sections 2.2, 4.1, and 7.1. A revised passage from "Defusing Ideological Defenses in Biology," published in *BioScience*, also contributes to section 7.1.

Figures

1.1 Bateman's (1948) data on male and female reproductive success in *Drosophila* 3
1.2 A recent representation of Bateman's principle 5
1.3 The Human Genome Project, 1000 Genomes Project, and Human Microbiome Project 13
2.1 A simple pattern in six objects, from Dennett (1991) 27
2.2 Map 27 of the Obesity System Atlas 37
2.3 Detail from Map 27 of the Obesity System Atlas 38
2.4 Map of the San Francisco Bay Area Rapid Transit system 51
3.1 Tidy relationship that may be posited of human aggression research 73
3.2 Alternative construal of human aggression research as uncovering causal patterns 78
6.1 Depiction of hierarchical levels of organization in an ecology textbook 167
6.2 A schema of mechanistic "levels" 183

Tables

1.1 Some of the questions neglected by genomic projects   14
2.1 Intertwined reasons to idealize   48
3.1 Payoffs for the prisoner's dilemma in one and multiple exchanges   64
4.1 Illustration of how aims influence scientific products   108
6.1 Articulation of distinct hierarchical concepts   181
7.1 Normative conclusions about scientific practice   219

Notes

Chapter One

1. Though there is a trend toward accounting for actual scientific practice, some philosophers do still explicitly pursue projects of rational reconstruction or accounts of an ultimate, completed science.

2. By "phenomena," I simply mean facts or occurrences in the world. I use this term throughout the book.

3. Of course, there is also much work that challenges this traditional dichotomy. For example, see Longino (2001) for a discussion of the dichotomy and an account of scientific objectivity that she thinks transcends it.

Chapter Two

1. To be clear, I have not here provided an argument against the existence or relevance of scientific laws. I use these well-known criticisms of universal, exceptionless laws simply to set the stage for the concept of a causal pattern. The latter may well be compatible with some accounts of laws.

2. In some cases, it may not be obvious whether deviation from a pattern constitutes an exception to a pattern or a limitation to its domain. Indeed, there may not always be a fact of the matter about which is so.

3. For defenses of the idea that physical phenomena are complex, see Cartwright (1983), Giere (1999), and Kennedy (2012).

4. On the other hand, such laws can be used to discover deviations and exceptions, and the deviations or exceptions uncovered can be of great scientific importance.

5. Woodward's account is one recent and especially popular version of a manipulability or interventionist theory of causation. Notice that, for Woodward, although causation is grounded in the concept of a manipulation, it does not actually require human action.

6. Cartwright (1983, 1989) also develops an account of causation that put human action center stage. She emphasizes that causal regularities must be posited to enact strategies effective at producing desired results. In contrast to Woodward, Cartwright's skepticism about laws leads

her to posit singular causes as basic, from which causal capacities derive. My views that uncovering causal patterns is central to science and that such patterns are real motivate the opposite prioritization.

7. This articulation of structural causes corresponds to Mackie's (1974) distinction between triggering causes and predisposing causes, where what I call structural causes are Mackie's predisposing causes. The idea also bears some similarity to Dretske's (2004) distinction between triggering and structuring causes. The difference is that, for Dretske, both triggering and structuring causes are parts of the causal process in question; that is, both refer to changes (at some stage) that were required to give rise to the effect in question.

8. Note that Woodward intends his generalizations about invariant causal relationships to be a successor to laws of nature. See especially Woodward (2000).

9. I suspect that, for similar reasons, a manipulability approach to causation is also conceptually basic, but I do not develop this idea here.

10. Mitchell couches this discussion as a criticism of Woodward's account of causation. However, she bases her remarks on Hausman and Woodward (1999). Woodward (2003) is explicit that modularity is a desirable property of representations of complex systems, not a property of causal systems themselves and not a requirement for causal attribution.

11. Wimsatt (1987) is republished as Chapter 6 in Wimsatt (2007). All page numbers in my citations refer to the later publication.

12. In this work I refer to "representations" that idealize. I mean this as a neutral term to include all of these scientific products and any other representational structures used in science. I set to the side questions of what exactly is being represented, as well as the nature of the representation relation. Accordingly, throughout the book, I either speak neutrally of the representation of phenomena or use Weisberg's (2013) terminology and speak in terms of representations of target system(s).

13. A version of this idea initially emerged from a discussion with Anthony Chemero, Thomas Polger, and Robert Skipper at the University of Cincinnati.

14. Adrian Currie helped me formulate the idea in this way.

15. This option was initially suggested to me by Guilherme Sanches de Oliveira.

16. The idea of representing as-if and its connection to fictions is reminiscent of the role that Vaihinger (1935) accorded to fictions, including in science. Some key differences are that Vaihinger claims all fictions are to be eventually eliminated, and he has in mind a much broader class of concepts and posits than just idealizations. See also Fine (1993) and Bokulich (2009). For more recent discussions of fictions and fictionalism in science, see the collection (Suárez 2009).

Chapter Three

1. Reciprocal altruism was first articulated by Trivers (1971). This is a very brief overview of reciprocal altruism and the prisoner's dilemma, but there is an expansive literature on each.

2. See Maynard Smith (1982); Potochnik (2012), and Rohwer and Rice (2013) for more extensive discussions of the idealizations involved in evolutionary game theory models, including the prisoner's dilemma.

3. Adrian Currie, in conversation, rightly points out that an additional or alternative reason for the lack of attention to specific evolutionary phenomena may be that we don't have fine-grain information about enough actual cases to develop specific models. This would be a different reason for an emphasis on general causal patterns.

Chapter Four

1. More specifically, according to this research, seeking and producing explanations furthers understanding in virtue of an emphasis on generalizations about broad patterns, and this same process can sometimes lead to overgeneralization. Here I set aside the connection between understanding and explanation to narrow the discussion, although the considerations raised here also have interesting implications for accounts of scientific explanation. I discuss this and related research in a bit more depth in section 4.2.

2. Winsberg argues that these model-building principles go beyond idealization and are best described as fictions; in Chapter 2, I briefly made the case for including Winsberg's fictions in the category of idealizations.

3. Because science is a social enterprise, some of these characteristics are, at least in part, collectively determined (Elihu Gerson, in conversation).

4. This fine-grained division into different deployments of individual aims is reminiscent of Kitcher's (2001) study of how research programs are shaped by the significance attributed to different questions.

5. In Chapter 5, I explicitly defend the idea that the formulation of scientific explanations is the systematic way to generate scientific understanding.

6. See Forster and Sober (1994) for an attempt to account for simplicity's contribution to truth in the special case of curve fitting.

Chapter Five

1. Aside from Craver and Strevens, an ontological approach to explanation has also been explicitly advocated by Salmon (1989).

2. This idea is closely related to Railton's (1981) notion of an "ideal explanatory text." According to Lewis, the difference is that on Lewis's view, this is a "vast structure" of causally related events, whereas Railton's ideal text consists of a long string of deductive-nomological arguments.

3. Douglas agrees with Trout (2002) that explanations cannot simply be valuable in virtue of generating a subjective sense of understanding in us. In Chapter 4, I distinguished between understanding as an epistemic accomplishment and the subjective sense of understanding.

4. Tellingly, an ideal gas law explanation can fail if a posited manipulability relationship does not obtain. For example, if a rigid container of a set volume is selected (and not because of its relationship to the other variables), then the gas's volume cannot be explained in terms of pressure, temperature, and amount of gas. In this case, no intervention on any other variable would change the gas's volume.

5. Strevens claims "causal ecumenism," but just as traditional ecumenism extends only to all Christian churches, Strevens's causal ecumenism extends only to physical accounts of causation.

6. Strevens (2008b) places much more emphasis on the positive potential of equilibrium explanations than on their limitations. In several pages emphasizing their value, he simply adds a parenthetical "(given that the different trajectories form a cohesive set)" (268). But, because of his reliance on physical causation, this requirement disqualifies many equilibrium explanations.

7. See Dretske (1972) for an early treatment of contrastive statements.

8. On this I am in agreement with van Fraassen (1980), who emphasizes the role of contrast classes but claims that contrast classes are only one way in which what he calls contextual factors influence explanations.

9. Some would attempt to characterize the difference here as a difference in explanandum by claiming that natural selection (represented in the game theory model) explains why the sparrows *evolved* this trait and genetics and environmental conditions explain why the sparrows *develop* the trait. But this is simply a way to direct attention to one part or another of the causal history of the same event, namely, that the sparrows have evolved to develop the trait in question. See also Potochnik (2010b).

10. Elsewhere, I propose a slightly different criterion for explanatory adequacy, [EA], intended to do this job (Potochnik 2015). I now think that criterion conflated the need for an explanation to account for an explanandum and some aspects of what it takes for a phenomenon to embody a causal pattern. Yet the criterion [EA] isn't sufficient to determine when a causal pattern is embodied. For these reasons, I don't employ the same criterion here.

11. To be precise, it is not a phenomenon itself but only ever some characterization of it that is explained.

12. Thanks to Adrian Currie for pushing me on some of these issues.

Chapter Six

1. This section and section 6.2.1 draw significantly from Potochnik and McGill (2012).

2. This may in part motivate the widespread idea that biological complexity has increased through evolutionary history (McShea 1991). I discuss how the conception of levels of organization relates to evolution just below.

3. Wimsatt (2007) is an exception. He endorses many of these theses about the significance of levels of organization, but he maintains that decomposability into discrete levels is not universal. Nonetheless, Wimsatt holds the view that "levels of organization are a deep, non-arbitrary, and extremely important feature of the ontological architecture of the natural world" (203).

4. Kim (2002) makes the same point, and Wimsatt (2007) argues that uniformity of composition fails at higher levels.

5. Polger and Shapiro (2016) argue that multiple realization is much less common than is customarily expected. But their view of realization does not provide the basis for discrete, well-articulated levels either.

6. I address the idea of mechanistic levels divorced from the classic conception of levels in the next section.

7. One might question this example on the basis that it seems that organisms and waste molecules are not on the same "level" either. But the waste molecules in question are independent components of the ecosystem. That is, they are not proper parts of any other whole that is itself a causally efficacious part of the ecosystem.

8. The sense of level intended here is also distinct from all those already under discussion. In this context, high-level causal assertions are simply those that do not reference the foundational, microphysical causal processes I am assuming here for the sake of argument. See below.

9. Differences in scale are, of course, not domain and interest relative; the relativity arises in where to demarcate levels.

Chapter Seven

1. Of course, to explain, a causal pattern must be embodied and its representation must be explanatorily adequate. See Chapter 5.

2. Quantitative genetics was introduced earlier, in the discussion of the scientific research into human aggression. This approach focuses on establishing the relative significance of genetics and environment in producing variation in a given trait.

3. Some proponents of evolutionary game theory, including notably Maynard Smith (1982), have views compatible with this construal.

4. See Potochnik (2013) for a more extensive treatment of this case.

5. The idea that physics is fully general is inaccurate of the actual content of physics; theories and models there are just as limited in scope and accuracy as elsewhere in science.

6. See Milam et al. (2011) for a more extended treatment of this feature of Roughgarden's research.

References

Achinstein, Peter. 1983. *The Nature of Explanation*. Oxford: Oxford University Press.
Akçay, Erol, Jeremy van Cleve, Marcus W. Feldman, and Joan Roughgarden. 2009. "A Theory for the Evolution of Other-Regard: Integrating Proximate and Ultimate Perspectives." *Proceedings of the National Academy of Sciences* 106:19061–19066.
Althoff, Robert R., and James J. Hudziak. 2011. "The Role of Behavioral Genetics in Child and Adolescent Psychiatry." *Journal of the Canadian Academy of Child and Adolescent Psychiatry* 20:4–5.
Amos, Christopher I., Margaret R. Spitz, and Paul Cinciripini. 2010. "Chipping Away at the Genetics of Smoking Behavior." *Nature Genetics* 42:366–368.
Atekwana, Estella A., William A. Sauck, and Douglas D. Werkema Jr. 2000. "Investigations of Geoelectrical Signatures at a Hydrocarbon Contaminated Site." *Journal of Applied Geophysics* 44:167–180.
Axelrod, Robert, and William D. Hamilton. 1981. "The Evolution of Cooperation." *Science* 211:1390–1396.
Bateman, A. J. 1948. "Intra-Sexual Selection in *Drosophila*." *Heredity* 2:349–368.
Batterman, Robert W. 2002a. "Asymptotics and the Role of Minimal Models." *British Journal for the Philosophy of Science* 53:21–38.
Batterman, Robert W. 2002b. *The Devil in the Details*. Oxford: Oxford University Press.
Batterman, Robert W. 2009. "Idealization and Modeling." *Synthese* 169:427–446.
Bechtel, William. 1984. "Reconceptualizations and Interfield Connections: The Discovery of the Link between Vitamins and Coenzymes." *Philosophy of Science* 51:265–292.
Bechtel, William. 2008. *Mental Mechanisms*. London: Routledge.
Bechtel, William, and Robert C. Richardson. 1993. *Discovering Complexity: Decomposition and Localization as Strategies in Scientific Research*. Princeton, NJ: Princeton University Press.
Bokulich, Alisa. 2008. *Reexamining the Quantum-Classical Relation: Beyond Reductionism and Pluralism*. Cambridge: Cambridge University Press.
Bokulich, Alisa. 2009. "Explanatory Fictions." In *Fictions in Science: Philosophical Essays on Modeling and Idealization*, edited by Mauricio Suárez, Routledge Studies in the Philosophy of Science, 91–109. New York: Routledge.

Bokulich, Alisa. 2013. "Explanatory Models Versus Predictive Models: Reduced Complexity Modeling in Geomorphology." In *EPSA11: Perspectives and Foundational Problems in Philosophy of Science*, edited by Vassilios Karakostas and Dennis Dieks, 115–128. Springer.

Bromberger, Silvain. 1966. "Why-Questions." In *Mind and Cosmos*, edited by R. Colodny, 86–111. Pittsburgh: University of Pittsburgh Press.

Brown, Joel S. 2001. "Fit of Form and Function, Diversity of Life, and Procession of Life as an Evolutionary Game." In *Adaptationism and Optimality*, edited by Steven Hecht Orzack and Elliott Sober, Cambridge Studies in Philosophy and Biology, chap. 4, 114–160. Cambridge: Cambridge University Press.

Campbell, Donald T. 1974. " 'Downward Causation' in Hierarchically Organised Biological Systems." In *Studies in the Philosophy of Biology*, edited by F. Ayala and T. Dobzhansky, 179–186. Berkeley: University of California Press.

Carnap, Rudolf. 1928. *The Logical Structure of the World*. Berkeley: University of California Press.

Cartwright, Nancy. 1983. *How the Laws of Physics Lie*. Oxford: Oxford University Press.

Cartwright, Nancy. 1989. *Nature's Capacities and Their Measurement*. Oxford: Clarendon Press.

Cartwright, Nancy. 1999. *The Dappled World: A Study of the Boundaries of Science*. Cambridge: Cambridge University Press.

Cartwright, Nancy. 2004. "Causation: One Word, Many Things." *Philosophy of Science* 71: 805–819.

Cartwright, Nancy. 2007. *Hunting Causes and Using Them: Approaches in Philosophy and Economics*. Cambridge: Cambridge University Press.

Cartwright, Nancy, Jordi Cat, Lola Fleck, and Thomas E. Uebel. 1996. *Otto Neurath: Philosophy Between Science and Politics*, vol. 38 of *Ideas in Context*. Cambridge: Cambridge University Press.

Chakravartty, Anjan. 2013. "Scientific Realism." In *The Stanford Encyclopedia of Philosophy*, edited by Edward N. Zalta. http://plato.stanford.edu/entries/scientific-realism/.

Chao, Hsiang-Ke, Szu-Ting Chen, and Roberta L. Millstein, eds. 2013. *Mechanism and Causality in Biology and Economics*. Springer.

Chemero, Anthony. 2011. *Radical Embodied Cognitive Science*. Cambridge, MA: MIT Press.

Chin-Parker, S., and A. Bradner. 2010. "Background Shifts Affect Explanatory Style: How a Pragmatic Theory of Explanation Accounts for Background Effects in the Generation of Explanations." *Cognitive Processing* 11:227–249.

Clark, James S. 1998. "Why Trees Migrate So Fast: Confronting Theory with Disperal Biology and the Paleorecord." *American Naturalist* 152:204–224.

Clutton-Brock, Tim. 2009. "Sexual Selection in Females." *Animal Behaviour* 77:3–11.

Cohen, L. Jonathan. 1992. *An Essay on Belief and Acceptance*. New York: Clarendon Press.

Corre, Valerie Le, Nathalie Machon, Remy J. Petit, and Antoine Kremer. 1997. "Colonization with Long-Distance Seed Dispersal and Genetic Structure of Maternally Inherited Genes in Forest Trees: A Simulation Study." *Genetical Research* 69:117–125.

Craig, Ian W., and Kelly E. Halton. 2009. "Genetics of Human Aggressive Behaviour." *Human Genetics* 126:101–113.

Craver, Carl F. 2006. "When Mechanistic Models Explain." *Synthese* 153:355–376.

Craver, Carl F. 2007. *Explaining the Brain: Mechanisms and the Mosaic Unity of Neuroscience*. Oxford: Oxford University Press.

Craver, Carl F. 2014. "The Ontic Conception of Scientific Explanation." In *Explanation in the Biological and Historical Sciences*, edited by Andreas Hütteman and Marie Kaiser. Springer.

Craver, Carl F., and William Bechtel. 2007. "Top-Down Causation Without Top-Down Causes." *Biology and Philosophy* 22:547–563.

Craver, Carl F., and Lindley Darden. 2005. "Introduction." *Studies in History and Philosophy of Biological and Biomedical Sciences* 36:233–244.

Curnow, R. N. and K. L. Ayres. 2007. "Population Genetic Models Can Be Used to Study the Evolution of the Interacting Behaviors of Parents and Their Progeny." *Theoretical Population Biology* 72:67–76.

Currie, Adrian. 2015. "Philosophy of Science and the Curse of the Case Study." In *The Palgrave Handbook of Philosophical Methods*, edited by Christopher Daly, chap. 21, 553–572. London: Palgrave Macmillan UK.

Darden, Lindley, and Nancy Maull. 1977. "Interfield Theories." *Philosophy of Science* 44:43–64.

Darwin, Charles. 1859. *On the Origin of Species by Means of Natural Selection, or the Preservation of Favoured Races in the Struggle for Life*. London: John Murray.

Darwin, Charles. 1871. *The Descent of Man and Selection in Relation to Sex*. London: John Murray.

de Regt, Henk W., and Dennis Dieks. 2005. "A Contextual Approach to Scientific Understanding." *Synthese* 144:137–170.

de Regt, Henk W., Sabina Leonelli, and Kai Eigner, eds. 2009. *Focusing on Scientific Understanding*. Pittsburgh: University of Pittsburgh Press.

Dennett, Daniel C. 1991. "Real Patterns." *The Journal of Philosophy* 88:27–51.

Douglas, Heather. 2009a. "Reintroducing Prediction to Explanation." *Philosophy of Science* 76:444–463.

Douglas, Heather. 2009b. *Science, Policy, and the Value-Free Ideal*. Pittsburgh, PA: University of Pittsburgh Press.

Dowe, Phil. 2000. *Physical Causation*. Cambridge: Cambridge University Press.

Dretske, Fred. 1972. "Contrastive Statements." *The Philosophical Review* 81:411–437.

Dretske, Fred. 2004. "Psychological vs. Biological Explanations of Behavior." *Behavior and Philosophy* 32:167–177.

Dupré, John. 1993. *The Disorder of Things: Metaphysical Foundations of the Disunity of Science*. Cambridge, MA: Harvard University Press.

Elgin, Catherine Z. 2004. "True Enough." *Philosophical Issues* 14:113–131.

Elgin, Catherine Z. 2007. "Understanding and the Facts." *Philosophical Studies* 132:33–42.

Elgin, Catherine Z. 2010. "Keeping Things in Perspective." *Philosophical Studies* 150:439–447.

Elliott, Kevin C. 2011. *Is a Little Pollution Good for You? Incorporating Societal Values in Environmental Research*. Oxford: Oxford University Press.

Elliott, Kevin C. 2013. "Douglas on Values: From Indirect Roles to Multiple Goals." *Studies in History and Philosophy of Science* 44:375–383.

Elliott, Kevin C., and Daniel J. McKaughan. 2014. "Nonepistemic Values and the Multiple Goals of Science." *Philosophy of Science* 81:1–21.

Elliott, Kevin C., and David Willmes. 2013. "Cognitive Attitudes and Values in Science." *Philosophy of Science* 80:807–817.

Epstein, Brian. 2012. "Agent-Based Modeling and the Fallacies of Individualism." In *Models, Simulations, and Representations*, edited by Paul Humphreys and Cyrille Imbert, chap. 6, 115–144. New York: Routledge.

Eronen, Markus I. 2013. "No Levels, No Problems: Downward Causation in Neuroscience." *Philosophy of Science* 80:1042–1052.

Eshel, Ilan, and Marcus W. Feldman. 2001. "Optimality and Evolutionary Stability under Short-Term and Long-Term Selection." In *Adaptationism and Optimality*, edited by Steven Hecht

Orzack and Elliott Sober, Cambridge Studies in Philosophy and Biology, chap. 4, 114–160. Cambridge: Cambridge University Press.

Fehr, Carla, and Kathryn S. Plaisance, eds. 2010. *Socially Relevant Philosophy of Science*, vol. 177 of *a Special Issue of Synthese*.

Feibleman, James K. 1954. "Theory of Integrative Levels." *The British Journal for the Philosophy of Science* 5:59–66.

Feynman, Richard. 1967. *The Character of Physical Law*. Cambridge, MA: MIT Press.

Fine, Arthur. 1993. "Fictionalism." *Midwest Studies in Philosophy* 18:1–18.

Fodor, Jerry. 1974. "Special Sciences: The Disunity of Science as a Working Hypothesis." *Synthese* 28:97–115.

Forster, Malcolm, and Elliott Sober. 1994. "How to Tell When Simpler, More Unified, or Less *Ad Hoc* Theories Will Provide More Accurate Predictions." *British Journal for the Philosophy of Science* 45:1–35.

Friedman, Michael. 1974. "Explanation and Scientific Understanding." *The Journal of Philosophy* 71:5–19.

Garfinkel, Alan. 1981. *Forms of Explanation: Rethinking the Questions in Social Theory*. New Haven, CT: Yale University Press.

Giere, Ronald N. 1988. *Explaining Science: A Cognitive Approach*. Chicago: University of Chicago Press.

Giere, Ronald N. 1999. *Science without Laws*. Chicago: University of Chicago Press.

Giere, Ronald N. 2004. "How Models Are Used to Represent Reality." *Philosophy of Science* 71:742–752.

Glenn, Andrea L., and Adrian Raine. 2014. "Neurocriminology: Implications for the Punishment, Prediction and Prevention of Criminal Behaviour." *Nature Reviews: Neuroscience* 15:54–63.

Godfrey-Smith, Peter. 2006. "The Strategy of Model-Based Science." *Biology and Philosophy* 21:725–740.

Godfrey-Smith, Peter, and Jon F. Wilkins. 2008. "Adaptationism." In *A Companion to the Philosophy of Biology*, edited by S. Sarkar and A. Plutynski, 186–201. Oxford: Wiley-Blackwell.

Gopnik, Alison. 1998. "Explanation as Orgasm." *Minds and Machines* 8:101–118.

Goss-Custard, J. D. 1977. "Predator Responses and Prey Mortality in the Redshank *Tringa totanus* (L.) and a Preferred Prey *Corophium volutator* (Pallas)." *Journal of Animal Ecology* 46:21–36.

Gould, Stephen Jay. 1996. *The Mismeasure of Man: Revised Edition*. New York: W. W. Norton.

Gould, Stephen Jay, and R. C. Lewontin. 1979. "The Spandrels of San Marco and the Panglossian Paradigm: A Critique of the Adaptationist Programme." *Proceedings of the Royal Society of London, Series B* 205:581–598.

Gowaty, Patricia Adair, Yong-Kyu Kim, and Wyatt W. Anderson. 2012. "No Evidence of Sexual Selection in a Repetition of Bateman's Classic Study of *Drosophila Melanogaster*." *Proceedings of the National Academy of Sciences* 109:11740–11745.

Grantham, Todd A. 2004. "Conceptualizing the (Dis)unity of Science." *Philosophy of Science* 71:133–155.

Grimm, Stephen R. 2006. "Is Understanding a Species of Knowledge?" *British Journal for the Philosophy of Science* 57:515–535.

Grimm, Stephen R. 2010. "The Goal of Explanation." *Studies in History and Philosophy of Science* 41:337–344.

Grimm, Stephen R. 2011. "Understanding." In *The Routledge Companion to Epistemology*, edited by Duncan Pritchard and Sven Berneker, 84–94. New York: Routledge.

Grimm, Stephen R. 2012. "The Value of Understanding." *Philosophy Compass* 7:103–117.
Guttman, Burton S. 1976. "Is 'Levels of Organization' a Useful Biological Concept?" *BioScience* 26:112–113.
Hartl, Daniel L., and Andrew G. Clark. 1997. *Principles of Population Genetics*. 3rd ed. Sunderland: Sinauer Associates.
Haug, Matthew C. 2010. "Realization, Determination, and Mechanisms." *Philosophical Studies* 150:313–330.
Hausman, Daniel, and James Woodward. 1999. "Independence, Invariance, and the Causal Markov Condition." *British Journal for the Philosophy of Science* 50:521–583.
Hempel, Carl. 1965. *Aspects of Scientific Explanation and Other Essays in the Philosophy of Science*. New York: Free Press.
Hempel, Carl. 1966. *Philosophy of Natural Science*. Englewood Cliffs, NJ: Prentice-Hall.
Hempel, Carl, and Paul Oppenheim. 1948. "Studies in the Logic of Explanation." *Philosophy of Science* 15:135–175.
Hesse, Mary. 1966. *Models and Analogies in Science*. Notre Dame, IN: University of Notre Dame Press.
Hilton, Denis J. 1990. "Conversational Processes and Causal Explanation." *Psychological Bulletin* 107:65–81.
Horgan, Terence E. 1982. "Supervenience and Microphysics." *Pacific Philosophical Quarterly* 63:29–43.
Hrdy, Sarah Blaffer. 1986. "Empathy, Polyandry, and the Myth of the Coy Female." In *Feminist Approaches to Science*, edited by R. Bleier, The Athene Series, 119–146. Oxford: Pergamon Press.
Huesmann, L. R., M. Moise-Titus, C. Podolski, and L. D. Eron. 2003. "Longitudinal Relations between Children's Exposure to TV Violence and Their Aggressive and Violent Behavior in Young Adulthood: 1977–1992." *Developmental Psychology* 39:201–221.
Hull, David L. 1978. "A Matter of Individuality." *Philosophy of Science* 45:335–360.
Isaac, Alistair M. C. 2013. "Modeling without Representation." *Synthese* 190:3611–3623.
Jackson, Frank, and Philip Pettit. 1992. "In Defense of Explanatory Ecumenism." *Economics and Philosophy* 8:1–21.
Jagers op Akkerhuis, Gerard A. J. M. 2008. "Analysing Hierarchy in the Organization of Biological and Physical Systems." *Biological Reviews* 83:1–12.
Johnson, Christopher N., Joanne L. Isaac, and Diana O. Fisher. 2007. "Rarity of a Top Predator Triggers Continent-Wide Collapse of Mammal Prey: Dingoes and Marsupials in Australia." *Proceedings of the Royal Society of London, Series B* 274:341–346.
Jones, Martin. 2005. "Idealization and Abstraction: A Framework." In *Idealization XII: Correcting the Model: Idealization and Abstraction in the Sciences*, edited by Martin Jones and Nancy Cartwright, 173–217. New York: Rodopi.
Kennedy, Ashley Graham. 2012. "A Non Representationalist View of Model Explanation." *Studies in History and Philosophy of Science* 43:326–332.
Kerr, Benjamin, and Peter Godfrey-Smith. 2002. "Individualist and Multi-Level Perspectives on Selection in Structured Populations." *Biology and Philosophy* 17:477–517.
Kim, Jaegwon. 1999. "Making Sense of Emergence." *Philosophical Studies* 95:3–36.
Kim, Jaegwon. 2002. "The Layered Model: Metaphysical Considerations." *Philosophical Explorations* 5:2–20.
Kimball, R. T., E. L. Braun, J. D. Ligon, V. Lucchini, and E. Randi. 2001. "A Molecular Phylogeny of the Peacock-Pheasants (Galliformes: *Polyplectron* spp.) Indicates Loss and Reduction of

Ornamental Traits and Display Behaviours." *Biological Journal of the Linnean Society* 73: 187–198.

Kitano, Hiroaki. 2002. "Systems Biology: A Brief Overview." *Science* 295:1662.

Kitcher, Philip. 1984. "1953 and All That: A Tale of Two Sciences." *Philosophical Review* 93:335–373.

Kitcher, Philip. 1992. "Gene: Current Usages." In *Keywords in Evolutionary Biology*, edited by Evelyn Fox Keller and Elisabeth A. Lloyd, 128–131. Cambridge, MA: Harvard University Press.

Kitcher, Philip. 2001. *Science, Truth, and Democracy*. Oxford: Oxford University Press.

Kitcher, Philip. 2011. *Science in a Democratic Society*. Amherst, NY: Prometheus Books.

Knight, Jonathan. 2002. "Sexual Stereotypes." *Nature* 415:254–256.

Krasnec, Michelle O., Chelsea N. Cook, and Michael D. Breed. 2012. "Mating Systems in Sexual Animals." *Nature Education Knowledge* 3:72.

Kvanvig, Jonathan. 2003. *The Value of Knowledge and the Pursuit of Understanding*. Cambridge: Cambridge University Press.

Ladyman, James, and Don Ross. 2007. *Every Thing Must Go*. Oxford: Oxford University Press.

Lande, Russell, and George F. Barrowclough. 1987. "Effective Population Size, Genetic Variation, and Their Use in Population Management." In *Viable Populations for Conservation*, edited by Michael E. Soulé, 87–123. Cambridge: Cambridge University Press.

Lange, Marc. 2013. "What Makes a Scientific Explanation Distinctively Mathematical?" *British Journal for the Philosophy of Science* 64:485–511.

Levins, Richard. 1966. "The Strategy of Model Building in Population Biology." *American Scientist* 54:421–431.

Levy, Arnon. 2014. "Machine-Likeness and Explanation by Decomposition." *Philosophers' Imprint* 14(6):1–15.

Lewis, David. 1986. "Causal Explanation." In *Philosophical Papers*, edited by David Lewis, vol. II, 214–240. Oxford: Oxford University Press.

Lidicker, William Z. Jr., 2008. "Levels of Organization in Biology: On the Nature and Nomenclature of Ecology's Fourth Level." *Biological Reviews* 83:71–78.

Lipton, Peter. 2009. "Understanding Without Explanation." In *Scientific Understanding: Philosophical Perspectives*, edited by Henk W. de Regt, Sabina Leonelli, and Kai Eigner, chap. 3, 43–63. Pittsburgh, PA: University of Pittsburgh Press.

Lloyd, Elisabeth A. 1988. *The Structure and Confirmation of Evolutionary Theory*. Princeton, NJ: Princeton University Press.

Lombrozo, Tania. 2011. "The Instrumental Value of Explanations." *Philosophy Compass* 6:539–551.

Lombrozo, Tania, and S. Carey. 2006. "Functional Explanation and the Function of Explanation." *Cognition* 99:167–204.

Longino, Helen E. 1990. *Science as Social Knowledge: Values and Objectivity in Scientific Inquiry*. Princeton, NJ: Princeton University Press.

Longino, Helen E. 2001. *The Fate of Knowledge*. Princeton, NJ: Princeton University Press.

Longino, Helen E. 2013. *Studying Human Behavior: How Scientists Investigate Aggression and Sexuality*. Chicago: University of Chicago Press.

MacArthur, Robert H. 1968. "The Theory of the Niche." In *Population Biology and Evolution*, edited by Richard C. Lewontin, 159–176. Syracuse, NY: Syracuse University Press.

Machamer, Peter, Lindley Darden, and Carl F. Craver. 2000. "Thinking about Mechanisms." *Philosophy of Science* 67:1–25.

Mackie, J. L. 1974. *The Cement of the Universe: A Study of Causation*. Oxford: Oxford University Press.

Maher, Brendan. 2008. "Personal Genomes: The Case of the Missing Heritability." *Nature* 456: 18–21.

Manuck, Stephen P., Janine D. Flory, Robert E. Ferrell, J. John Mann, and Matthew F. Muldoon. 2000. "A Regulatory Polymorphism of the Monoamine Oxidase-A Gene May Be Associated with Variability in Aggression, Impulsivity, and Central Nervous System Serotonergic Responsivity." *Psychiatry Research* 95:9–23.

May, R. M. 1976. "Models for Two Interacting Populations." In *Theoretical Ecology: Principles and Applications*, edited by R. M. May, 49–70. Philadelphia: W. B. Saunders.

Maynard Smith, John. 1982. *Evolution and the Theory of Games*. Cambridge: Cambridge University Press.

Maynard Smith, John, and Eörs Szathmáry. 1995. *The Major Transitions in Evolution*. Oxford: Oxford University Press.

McGill, Brian J. 2010. "Ecology: Matters of Scale." *Science* 328:575.

McMullin, Ernan. 1985. "Galilean Idealization." *Studies in History and Philosophy of Science* 16:247–273.

McShea, Daniel W. 1991. "Complexity and Evolution: What Everybody Knows." *Biology and Philosophy* 6:303–324.

Melnyk, Andrew. 2013. "Can Metaphysics Be Naturalized? And If So, How?" In *Scientific Metaphysics*, edited by Don Ross, James Ladyman, and Harold Kincaid, 79–95. Oxford: Oxford University Press.

Milam, Erika, Roberta Millstein, Angela Potochnik, and Joan Roughgarden. 2011. "Sex and Sensibility: The Role of Social Selection, a Review Symposium of Roughgarden's *The Genial Gene*." *Metascience* 20:253–277.

Mitchell, Sandra D. 2003. *Biological Complexity and Integrative Pluralism*. Cambridge Studies in Philosophy and Biology. Cambridge: Cambridge University Press.

Mitchell, Sandra D. 2012a. "Emergence: Logical, Functional and Dynamical." *Synthese* 185:171–186.

Mitchell, Sandra D. 2012b. *Unsimple Truths: Science, Complexity, and Policy*. Chicago: University of Chicago Press.

Mitchell, William A., and Thomas J. Valone. 1990. "The Optimization Research Program: Studying Adaptations by Their Function." *The Quarterly Review of Biology* 65:43–52.

Molles, Manuel C. 2002. *Ecology: Concepts and Applications*. New York: McGraw-Hill.

Morrison, Margaret, and Mary S. Morgan. 1999. "Models as Mediating Instruments." In *Models as Mediators*, edited by Mary S. Morgan and Margaret Morrison, chap. 1, 1–9. Cambridge: Cambridge University Press.

Nagel, Ernest. 1961. *The Structure of Science*. London: Routledge and Kegen Paul.

Neurath, Otto. 1936a. "Encyclopedia as 'Model.'" In *Philosophical Papers 1913–1946*, edited by Robert S. Cohen and Marie Neurath, vol. 16 of *Vienna Circle Collection*. Dordrecht: D. Reidel.

Neurath, Otto. 1936b. "Individual Sciences, Unified Science, Pseudo-Rationalism." In *Philosophical Papers 1913–1946*, edited by Robert S. Cohen and Marie Neurath, vol. 16 of *Vienna Circle Collection*. Dordrecht: D. Reidel.

Neurath, Otto. 1937. "Unified Science and its Encyclopedia." In *Philosophical Papers 1913–1946*, edited by Robert S. Cohen and Marie Neurath, vol. 16 of *Vienna Circle Collection*. Dordrecht: D. Reidel.

Neurath, Otto. 1983. *Philosophical Papers 1913–1946*, vol. 16 of *Vienna Circle Collection*. Dordrecht: D. Reidel.

Ney, Alyssa. 2009. "Physical Causation and Difference-Making." *The British Journal for the Philosophy of Science* 60:737–764.

Odenbaugh, Jay, and Anna Alexandrova. 2011. "Buyer Beware: Robustness Analyses in Economics and Biology." *Biology and Philosophy* 26:757–771.

Okasha, Samir. 2006. *Evolution and the Levels of Selection*. Oxford: Oxford University Press.

O'Neill, R. V., D. L. DeAngelis, J. B. Waide, and T. F. H. Allen. 1986. *A Hierarchical Concept of Ecosystems*. Princeton, NJ: Princeton University Press.

Oppenheim, Paul, and Hilary Putnam. 1958. "Unity of Science as a Working Hypothesis." In *Minnesota Studies in the Philosophy of Science*, edited by Herbert Feigl, Michael Scriven, and Grover Maxwell, vol. 2, 3–36. Minneapolis: University of Minnesota Press.

Oreskes, Naomi, and Erik M. M. Conway. 2011. *Merchants of Doubt: How a Handful of Scientists Obscured the Truth on Issues from Tobacco Smoke to Global Warming*. New York: Bloomsbury Press.

Orzack, Steven Hecht, and Elliott Sober. 1993. "A Critical Assessment of Levins's 'The Strategy of Model Building in Population Biology' (1966)." *Quarterly Review of Biology* 68:533–546.

Orzack, Steven Hecht, and Elliott Sober. 1994. "Optimality Models and the Test of Adaptationism." *The American Naturalist* 143:361–380.

Owens, I. P. F. and R. V. Short. 1995. "Hormonal Basis of Sexual Dimorphism in Birds: Implications for New Theories of Sexual Selection." *Trends in Ecology and Evolution* 10:44–47.

Oyama, Susan, Russell D. Gray, and Paul E. Griffiths, eds. 2001. *Cycles of Contingency: Developmental Systems and Evolution*. Life and Mind: Philosophical Issues in Biology and Psychology, Boston: MIT Press.

Parker, Wendy S. 2011. "When Climate Models Agree: The Significance of Robust Model Predictions." *Philosophy of Science* 78:579–600.

Pavlov, Konstantin A., Dimitry A. Chistiakov, and Vladimir P. Chekhonin. 2012. "Genetic Determinants of Aggression and Impulsivity in Humans." *Journal of Applied Genetics* 53:61–82.

Paynter, Nina P., Daniel I. Chaseman, Guillaume Paré, Julie E. Buring, Nancy R. Cook, Joseph P. Miletich, and Paul M. Ridker. 2010. "Association between a Literature-Based Genetic Risk Score and Cardiovascular Events in Women." *The Journal of the American Medical Association* 303:631–637.

Piccinini, Gualtiero, and Corey J. Maley. 2014. "The Metaphysics of Mind and the Multiple Sources of Multiple Realizability." In *New Waves in Philosophy of Mind*, edited by Mark Sprevak and Jesper Kallestrup, 125–152. London: Palgrave Macmillan UK.

Pincock, Christopher. 2012. *Mathematics and Scientific Representation*. Oxford: Oxford University Press.

Polger, Thomas W. 2010. "Mechanisms and Explanatory Realization Relations." *Synthese* 177:193–212.

Polger, Thomas W., and Lawrence A. Shapiro. 2016. *The Multiple Realization Book*. Oxford: Oxford University Press.

Potochnik, Angela. 2009. "Optimality Modeling in a Suboptimal World." *Biology and Philosophy* 24:183–197.

Potochnik, Angela. 2010a. "Explanatory Independence and Epistemic Interdependence: A Case Study of the Optimality Approach." *The British Journal for the Philosophy of Science* 61:213–233.

Potochnik, Angela. 2010b. "Levels of Explanation Reconceived." *Philosophy of Science* 77:59–72.

Potochnik, Angela. 2011a. "Explanation and Understanding: An Alternative to Strevens' *Depth*." *European Journal for the Philosophy of Science* 1:29–38.

Potochnik, Angela. 2011b. "A Neurathian Conception of the Unity of Science." *Erkenntnis* 34: 305–319.

Potochnik, Angela. 2012. "Modeling Social and Evolutionary Games." *Studies in History and Philosophy of Biological and Biomedical Sciences* 43:202–208.

Potochnik, Angela. 2013. "Defusing Ideological Defenses in Biology." *BioScience* 63:118–123.

Potochnik, Angela, ed. 2014. *Socially Engaged Philosophy of Science*, vol. 79 of *A Special Issue of Erkenntnis*.

Potochnik, Angela. 2015. "Causal Patterns and Adequate Explanations." *Philosophical Studies* 172:1163–1182.

Potochnik, Angela. 2016. "Scientific Explanation: Putting Communication First." *Philosophy of Science* 83:721–732.

Potochnik, Angela, and Brian McGill. 2012. "The Limitations of Hierarchical Organization." *Philosophy of Science* 79:120–140.

Putnam, Hilary. 1975. *Philosophy and our Mental Life*, vol. 2 of *Philosophical Papers*, chap. 14, 291–303. Cambridge: Cambridge University Press.

Railton, Peter. 1981. "Probability, Explanation, and Information." *Synthese* 48:233–256.

Rice, Collin. 2015. "Moving Beyond Causes: Optimality Models and Scientific Explanation." *Noûs* 49:589–615.

Ricklefs, Robert E. 2008. "Disintegration of the Ecological Community." *The American Naturalist* 172:741–750.

Rohwer, S., and P. W. Ewald. 1981. "The Cost of Dominance and Advantage of Subordinance in a Badge Signalling System." *Evolution* 35:441–454.

Rohwer, Yasha, and Collin Rice. 2013. "Hypothetical Pattern Idealization and Explanatory Models." *Philosophy of Science* 80:334–355.

Rosenberg, Alexander. 1994. *Instrumental Biology or the Disunity of Science*. Chicago: University of Chicago Press.

Roughgarden, Joan. 2004. *Evolution's Rainbow: Diversity, Gender, and Sexuality in Nature and People*. Berkeley: University of California Press.

Roughgarden, Joan. 2009. *The Genial Gene: Deconstructing Darwinian Selfishness*. Berkeley: University of California Press.

Rueger, Alexander, and Patrick McGivern. 2010. "Hierarchies and Levels of Reality." *Synthese* 176:379–397.

Sadava, David E., David M. Hillis, H. Craig Heller, and May Berenbaum. 2009. *Life: The Science of Biology*. 9th ed. W. H. Freeman.

Salmon, Wesley. 1984. *Scientific Explanation and the Causal Structure of the World*. Princeton, NJ: Princeton University Press.

Salmon, Wesley. 1989. *Four Decades of Scientific Explanation*. Minneapolis: University of Minnesota Press.

Salmon, Wesley C. 1978. "Why Ask, 'Why?'? An Inquiry Concerning Scientific Explanation." *Proceedings and Addresses of the American Philosophical Association* 51:683–705.

Sazima, Ivan, Lucelia Nobre Carvalho, Fernando Pereira Mendonca, and Jansen Zuanon. 2006. "Fallen Leaves on the Water-Bed: Diurnal Camouflage of Three Night Active Fish Species in an Amazonian Streamlet." *Neotropical Ichthyology* 4:119–122.

Schoener, Thomas W. 1986. "Mechanistic Approaches to Community Ecology: A New Reductionism." *American Zoologist* 26:81–106.

Shapiro, Lawrence A. 2000. "Multiple Realizations." *Journal of Philosophy* 97:635–654.
Skipper, Robert A., and Roberta L. Millstein. 2005. "Thinking about Evolutionary Mechanisms: Natural Selection." *Studies in History and Philosophy of Biological and Biomedical Sciences* 36:327–347.
Sober, Elliott. 1983. "Equilibrium Explanation." *Philosophical Studies* 43:201–210.
Sober, Elliott. 1999. "The Multiple Realizability Argument against Reduction." *Philosophy of Science* 66:542–564.
Sober, Elliott, and David Sloan Wilson. 1998. *Unto Others: The Evolution and Psychology of Unselfish Behavior*. Cambridge, MA: Harvard University Press.
Sterelny, Kim. 1996. "Explanatory Pluralism in Evolutionary Biology." *Biology and Philosophy* 11:193–214.
Stevens, Martin, and Sami Merilaita. 2009. "Animal Camouflage: Current Issues and New Perspectives." *Philosophical Transactions of the Royal Society B* 364:423–427.
Strevens, Michael. 2004. "The Causal and Unification Accounts of Explanation Unified—Causally." *Nous* 38:154–179.
Strevens, Michael. 2006. *Bigger Than Chaos: Understanding Complexity through Probability*. Cambridge, MA: Harvard University Press.
Strevens, Michael. 2007. "Review of Woodward, *Making Things Happen*." *Philosophy and Phenomenological Research* 74:233–249.
Strevens, Michael. 2008a. "Comments on Woodward, *Making Things Happen*." *Philosophy and Phenomenological Research* 77:171–192.
Strevens, Michael. 2008b. *Depth: An Account of Scientific Explanation*. Cambridge, MA: Harvard University Press.
Strevens, Michael. 2013. "No Understanding without Explanation." *Studies in History and Philosophy of Science* 44:510–515.
Suárez, Mauricio, ed. 2009. *Fictions in Science: Philosophical Essays on Modeling and Idealization*. Routledge Studies in the Philosophy of Science. New York: Routledge.
Suppes, Patrick. 2002. *Representation and Invariance of Scientific Structures*. Stanford, CA: CSLI Publications.
Tabery, James. 2014. *Beyond Versus: The Struggle to Understand the Interaction of Nature and Nurture*. Life and Mind: Philosophical Issues in Biology and Psychology. Cambridge, MA: MIT Press.
Takahashi, Mariko, Hiroyuki Arita, Mariko Hiraiwa-Hasegawa, and Toshikazu Hasegawa. 2008. "Peahens Do Not Prefer Peacocks with More Elaborate Trains." *Animal Behaviour* 75:1209–1219.
Trivers, Robert L. 1971. "The Evolution of Reciprocal Altruism." *The Quarterly Review of Biology* 46:35–57.
Trout, J. D. 2002. "Scientific Explanation and the Sense of Understanding." *Philosophy of Science* 69:212–233.
Trout, J. D. 2007. "The Psychology of Scientific Explanation." *Philosophy Compass* 2:564–591.
Vaihinger, Hans. 1935. *The Philosophy of 'As If': A System of the Theoretical, Practical and Religious Fictions of Mankind*. London: Routledge and Kegan Paul.
van Fraassen, Bas C. 1980. *The Scientific Image*. Oxford: Clarendon Press.
van Fraassen, Bas C. 2008. *Scientific Representation: Paradoxes of Perspective*. Oxford: Oxford University Press.
Vandenbroeck, Ir Philippe, Jo Goossens, and Marshall Clemens. 2007. "Tackling Obesities: Future Choices—Obesity System Atlas." http://www.foresight.gov.uk

Waters, C. Kenneth. 1990. "Why the Anti-Reductionist Consensus Won't Survive: The Case of Classical Mendelian Genetics." *Philosophy of Science Association* 1:125–139.

Weisberg, Michael. 2006. "Robustness Analysis." *Philosophy of Science* 73:730–742.

Weisberg, Michael. 2007. "Three Kinds of Idealization." *The Journal of Philosophy* 104:639–659.

Weisberg, Michael. 2013. *Simulation and Similarity: Using Models to Understand the World*. Oxford: Oxford University Press.

West, Geoffrey B., and James H. Brown. 2005. "The Origin of Allometric Scaling Laws in Biology from Genomes to Ecosystems: Towards a Quantitative Unifying Theory of Biological Structure and Organization." *Journal of Experimental Biology* 208:1575–1592.

Wilkinson, Gerald S. 1984. "Reciprocal Food Sharing in the Vampire Bat." *Nature* 308:181–184.

Williams, Joseph J., Tania Lombrozo, and Bob Rehnder. 2013. "The Hazards of Explanation: Overgeneralization in the Face of Exceptions." *Journal of Experimental Psychology: General* 142:1006–1014.

Wilson, Edward O., and William H. Bossert. 1971. *A Primer of Population Biology*. Sunderland: Sinauer Associates.

Wilson, Robert A. 2008. "The Biological Notion of Individual." In *The Stanford Encyclopedia of Philosophy*, edited by Edward N. Zalta. http://plato.stanford.edu/archives/fall2008/entries/biology-individual

Wimsatt, William C. 1981. "Robustness, Reliability, and Overdetermination." In *Scientific Inquiry and the Social Sciences*, edited by M. B. Brewer and B. E. Collins, 124–163. San Francisco: Jossey-Bass.

Wimsatt, William C. 1987. "False Models as Means to Truer Theories." In *Neutral Models in Biology*, edited by N. Nitecki and A. Hoffman, 23–55. Oxford: Oxford University Press.

Wimsatt, William C. 2007. *Re-Engineering Philosophy for Limited Beings*. Cambridge, MA: Harvard University Press.

Winsberg, Eric B. 2010. *Science in the Age of Computer Simulation*. Chicago: University of Chicago Press.

Woodward, James. 2000. "Explanation and Invariance in the Special Sciences." *British Journal for the Philosophy of Science* 51:197–254.

Woodward, James. 2003. *Making Things Happen: A Theory of Causal Explanation*. Oxford: Oxford University Press.

Woodward, James. 2007. "Causation with a Human Face." In *Causation, Physics, and the Constitution of Reality: Russell's Republic Revisited*, edited by Huw Price and Richard Corry, 66–105. Oxford: Oxford University Press.

Woodward, James. 2008. "Response to Strevens." *Philosophy and Phenomenological Research* 77:193–212.

Worden, Lee, and Simon A. Levin. 2007. "Evolutionary Escape from the Prisoner's Dilemma." *Journal of Theoretical Biology* 245:411–422.

Wright, Cory D. 2012. "Mechanistic Explanation without the Ontic Conception." *European Journal for the Philosophy of Science* 2:375–394.

Wylie, Alison. 1999. "Rethinking Unity as a 'Working Hypothesis' for Philosophy of Science: How Archaeologists Exploit the Disunities of Science." *Perspectives on Science* 7:293–317.

Zagzebski, Linda. 2001. "Recovering Understanding." In *Knowledge, Truth, and Duty: Essays on Epistemic Justification, Responsibility, and Virtue*, edited by Matthias Steup, 235–252. Oxford: Oxford University Press.

Index

abstraction, 6, 41, 43, 55
acceptance, 97, 105, 192. *See also* epistemic acceptance
Achinstein, Peter, 126
adaptationism, 10, 69, 201, 202
aggression, 17, 62, 63, 70–80, 84, 87, 88, 110, 203, 210. *See also* sex differences: aggression and promiscuity
aims of science, 2, 11, 16, 19, 21, 60, 62, 65, 67, 69, 80, 86, 88, 91, 92, **90–93**, 97–98, 101, 102, 104–105, **105–112**, 113, 117, 134, 153, 161, 186, 190, 191, 194, 195, 197–199, 204, 206–208, 210, 213–217, 219–221
  deployments of aims, 107, 108, 110, 111, 153, 201, 213
Akçay, Erol, 65
Alexandrova, Anna, 92
Allen, T. F. H., 168
Althoff, Robert R., 71
altruism, reciprocal. *See* cooperation
Amos, Christopher I., 154, 155
analogies and scientific modeling, 53
asexual reproduction, assumption of, 64–66, 68, 96
astrology, 93, 116, 215
astronomy, 187
Atekwana, Estella A., 196
Axelrod, Robert, 63, 66
Ayres, K. L., 66

background knowledge, 87, 92, 100, 147, 156, 203
Barrowclough, George F., 56
Bateman, A. J., 3–5
Bateman's principle. *See* sex differences: Bateman's principle
Batterman, Robert W., 44, 80–81, 103, 138

BDNF gene, 153, 156
Bechtel, William, 2, 164, 165, 177, 181, 182, 194
behavioral ecology, 19, 62–70, 78, 79, 81, 86, 90, 101, 111, 142
behavioral genetics, 70–71, 73, 74, 216. *See also* genetics
biogeophysics, 187
Bokulich, Alisa, 56, 82–84, 109, 121, 143, 230
Bossert, William H., 168
Boyle's law. *See* ideal gas law
Bradner, A., 149
Bromberger, Silvain, 126, 135
Brown, James H., 168
Brown, Joel S., 201

Campbell, Donald T., 175
Carey, Susan, 113
Carnap, Rudolf, 1. *See also* philosophy of science, approaches to: rational reconstruction of science
Cartwright, Nancy, 2, 16, 24–25, 28, 33, 41, 43, 55, 106, 188, 214–215, 229
causal complexity, 2, 19, 21, 23, **35–41**, 41, 45, 47, 57, 62, 71, 74, 76, 85, 86, 88, 114, 127, 134, 148, 153, 168–171, 175, 176, 184, 186, 187, 190, 194–197, 200, 214, 217, 221
causal pattern explanation, **135–145**, 145
causal patterns, 19, 23, **25–33**, 34, 41, 45, 57, 60, 62, 67–71, 76, 77, 79–81, 85–86, 88, 89, 91, 97, 101, 114, 115, 118–120, 134, 136, 137, 139, 143, 144, 149, 153, 155, 160, 162, 172, 182, 183, 191, 193, 206, 208, 213, 215, 217–218, 220
  embodied, 20, 26–29, 42, 43, 45, 55, 62, 78, 95, 144, 148, 153, 156
causation, 30, 33–34, 39, 49, 67, 196, 202, 207, 215 (*see also* scope of causal dependence)

causation (*continued*)
  causal foundationalism, 178
  direction of, 162
  manipulationist approach, 19, 23, 29–35, 40, 83, 114, 136, 138, 175, 178
  physical account of, 31, 137, 138, 141, 143
Chakravartty, Anjan, 119
Chao, Hsiang-Ke, 9
Chemero, Anthony, 15, 230
Chen, Szu-Ting, 9
Chin-Parker, S., 149
Clark, Andrew G., 53
Clark, James S., 184
classical mechanics, 82, 83, 90, 99, 100, 109, 121, 210
climate change, 14, 84, 86, 184, 204, 221
climate science, 5, 19, 62, 84–88, 90, 106, 111, 204
closed orbit theory. *See* classical mechanics
Clutton-Brock, Tim, 6
cognitive attitudes, 96, 105
Cohen, L. Jonathan, 96
common sense, 210–212
complexity, 14–15, 18, 45, 47, 48, 58, 85, 108, 118, 129, 162, 164, 168, 193
  in biological phenomena, 4, 12, 14, 16, 17, 35–40, 69, 71, 74, 75, 163–164, 172, 176
  causal (*see* causal complexity)
  and human limitations, 15, 45, 47, 58, 108, 111, 134, 188
  and policy decisions, 16
  and the scientific enterprise, 1–2, 16, 18, 127
complex realization, 172, 173, 180. *See also* realization and multiple realization
composition, 33, 162, 165, 166, 168–171, 173–179, 183–186, 208, 221
  mechanistic, 178–183
computational models, 84, 85, 87
contextual empiricism, 216
Conway, Eric, 5
cooperation, 62–70, 77, 78, 81, 84, 87, 88, 90, 107, 142, 203, 210, 215, 220, 221
coordinate unity, 196
Corre, Valerie Le, 184
cosmology, 187
Craig, Ian W., 71, 74, 77
Craver, Carl F., 9, 25, 123–125, 135, 164, 165, 177, 181, 182, 194
Curnow, R. N., 66
Currie, Adrian, 9, 10, 188, 230, 232

Darden, Lindley, 9, 25, 135, 164, 182, 194
Darwin, Charles, 2, 4, 6, 212
DeAngelis, D. L., 168
de-idealization, 44, 48, 58, 66, 87, 90, 92
Dennett, Daniel C., 26–29, 115
de Regt, Henk W., 9, 131

determinism, 209
developmental biology, 14, 40, 150
Dieks, Dennis, 9
difference-making, 33, 135, 141, 145
disagreement in science, 20, 21, 110
division of labor in science, 73–79, 162, 185, 188. *See also* interdisciplinary collaboration; unity of science
Douglas, Heather, 105, 109, 129
Dowe, Phil, 31, 33
Dretske, Fred, 230, 231
drift, 201
Dupré, John, 2, 10–11, 16, 188, 192, 209–210, 214

effective population size, 53, 54, 56, 59
Elgin, Catherine Z., 7, 20, 93–99, 103, 105, 116
Elliott, Kevin C., 96, 98, 105, 199, 205
emergence, 161, 163–164, 169
empirical confirmation, 67–68, 78–79, 81, 87–88, 92, 109
environmental issues. *See* climate science
epigenetics, 12, 15, 36, 74, 107, 111, 151
epistasis, 74
epistemic acceptability, 95–98, **99–100**, 109, 113, 116, 117, 192, 200, 204, 206
epistemic acceptance, 97, 100, 105
epistemic interdependence, 190, 191, 194–196, 198, 218
epistemic values, 120
Epstein, Brian, 168, 172, 221
equilibrium models, 31–32, 49, 194
Eronen, Markus I., 170, 175, 177, 179, 182, 183, 185
Eshel, Ilan, 67, 68, 194
evidential interdependence. *See* epistemic interdependence
evolution and mate selection. *See* sexual selection
evolutionary biology. *See* population biology
evolutionary game theory, 31, 49, 55, 63–69, 107, 137, 139, 142, 150, 157, 159, 194, 200–202, 215, 220, 232
Ewald, P. W., 150
explanation, 2, 9, 10, 20, 43, 92, 96, 105–107, 109, 120, 132, 133, 153, **122–160**, 161, 165, 189, 195, 198, 204, 208, 217, 218 (*see also* understanding; why-questions)
  and accuracy, 24, 26, 132, 134, 149 (*see also* representation: and accuracy)
  causal approach, 125, 129, 135
  causal-pattern approach, 20 (*see also* causal patterns)
  cognitive features, 122–123, 125, 128, 130, 131, 133
  communicative approach, 147, 153
  communicative role, 123, 125, 128–131, 134, 136, 218
  contrastive, 146

INDEX
249

deductive-nomological approach, 122, 125, 130, 155
equilibrium, 137–140, 157
and exceptions, 155
and expectability, 122, 136, 141, 155, 159
explanatory adequacy, 20, 134, 154–160, 218
explanatory asymmetry, 135
and generality, 122, 136, 141
and idealization (*see* idealization)
inference to the best explanation, 109
kairetic account, 17–18, 135, 141, 142
mechanistic approach, 125
minor vs. main causal factors, 155
non-causal, 138, 139, 143
ontic account, 123–125, 130, 133, 135, 148
pragmatic account, 126, 149
psychological requirements (*see* explanation: cognitive features)
reductive vs. high-level (*see* levels of organization: and scientific explanation)
role in science, 124
and understanding, 113, 122–124, 129, **133**

Fagan, Melinda, 171
Fehr, Carla, 6
Feibleman, James K., 163–164, 168
Feldman, Marcus W., 65, 67, 68, 194
Feynman, Richard, 24
Fine, Arthur, 230
fitness, 63–65
fluid dynamics, 19, 62, 101
Fodor, Jerry, 165, 173, 179, 186, 188
Forster, Malcolm, 231
frictionless plane, 2, 52

game theory. *See* evolutionary game theory
Garfinkel, Alan, 122, 141, 146, 155, 165, 173
genetic drift, 36, 49, 53–55, 60, 65, 99, 101, 158, 159, 193
genetics, 10, 36, 40, 49, 66, 68, 69, 71, 107, 139, 153, 155, 157, 159, 176, 192, 202 (*see also* behavioral genetics; molecular genetics; population genetics)
genetic-environmental interaction, 77, 151, 152
genomics, 16, 17, 35
1000 Genomes Project, 12–13, 35
Human Genome Project, 11–13, 35
Human Microbiome Project, 12–13, 35
and IQ differences, 5
geology, 187, 212
Gerson, Elihu, 231
Giere, Ronald N., 24, 25, 53, 103, 229
Godfrey-Smith, Peter, 56, 65, 69, 194
Gopnik, Alison, 113–114
Goss-Custard, J. D., 31
Gould, Stephen Jay, 4, 69, 201

Gowaty, Patricia Adair, 4
Grantham, Todd A., 194
grasping, 94, 114, 116, 130, 131
gravitation, law of universal, 24–26, 29, 43, 51
Gray, Russell D., 15
Griesemer, James, 194
Griffiths, Paul E., 15
Grimm, Stephen R., 94, 103, 114, 130, 132
Guttman, Burton S., 171, 175

Halton, Kelly E., 71, 74, 77
Hamilton, William D., 63, 66
Harris sparrows, 150, 157
Hartl, Daniel L., 53
Haug, Matthew C., 179, 182
Hausman, Daniel, 230
hawk-dove game, 66, 150, 157, 158
Heller, H. Craig, 166
Hempel, Carl, 25, 122, 130, 165, 173, 188
Hesse, Mary, 53
hierarchy. *See* levels of organization
Hillis, David, 166
Hilton, Denis J., 149
Horgan, Terence E., 164
Hrdy, Sarah Blaffer, 5–6, 203
Hudziak, James J., 71
Huesmann, L. R., 72
Hull, David L., 10

ideal gas law, 26, 28, 29, 31, 32, 42, 43, 98, 137, 139, 144, 155
idealization, 2, 6, 18–19, 21, 23, 41, 43, 44, **41–61**, 62, 64–66, 69, 70, 74–76, 79–81, 87–90, 93–94, 97, 101, 102, 105, 106, 108, 110, 111, 116, 118, 119, 129, 132, 144, 153, 155, 157, 162, 172, 175, 176, 186, 190, 192
and computational limits, 103
and disciplinary differences, 70, 194, 195, 201, 202, 213, 219, 220 (*see also* de-idealization)
as fiction, 55–57, 83, 143
intertwined reasons for, 23, 47, 58–60, 75, 91, 103–104
pedagogical value, 103
rampant and unchecked, 23, 91, 93, 97, 134, 161, 199, 213, 215, 217, 218, 221
representational role (*see* idealization: representing as-if)
representing as-if, 19, 23, 41, **52**, 52, 53, 55, 57, 59, 60, 62, 68, 83, 91, 95, 97
incommensurability, 216
infinite population, assumption of, 49, 52, 55, 60, 64, 65, 96, 99–101, 193, 201
integrated models, 193–194
interdisciplinary collaboration, 194, 196, 220, 221
intervention, 30–31, 34, 40, 141, 229
invariance, 30, 39, 40, 141

Isaac, Alistair M. C., 106
isomorphism. *See* model-target relationship

Jackson, Frank, 122, 141
Jagers op Akkerhuis, Gerard A. J. M., 168

Kelvin, Lord (William Thomson), 212
Kennedy, Ashley Graham, 229
Kerr, Benjamin, 65, 69
Kim, Jaegwon, 164, 170, 174, 232
Kimball, R. T., 190
Kitano, Hiroaki, 15
Kitcher, Philip, 9, 192, 205, 231
Knight, Jonathan, 4
knowledge, 20, 91, 119, 141, 204, 206
Krasnec, Michelle O., 4
Kvanvig, Jonathan, 102

Ladyman, James, 166, 210–212
Lande, Russell, 56
Lange, Marc, 143
laws, 2, 18, 24–25, 29, 33, 34, 43, 51, 89, 106, 135, 188, 214
levels of organization, 2, 11, 20, 161, 162, 170, 174, 176, 178, 180, 185, 187, 188, 207, 214, 217, 218, 220–221
    and abstractness, 180
    in biology, 9, 166, 168, 169
    compositional hierarchy (*see* composition)
    and evolution, 169
    metaphysical vs. scientific, 165–166, 169–170, 186, 187, 195
    in physics, 9, 168
    and properties, 11, 163–165, 169, 171–174
    quasi levels, 184–185
    and scientific disciplines, 21, 162, 175, 186–187, 221
    and scientific explanation, 141, 165, 173, 174
    in the social sciences, 168
levels of selection. *See* natural selection
Levin, Simon A., 65
Levins, Richard, 46, 47, 107
Lewis, David, 107, 151
Lewontin, R. C., 69, 201
Lidicker, William Z., Jr., 168
Lipton, Peter, 132
Lloyd, Elisabeth A., 67
logical empiricism, 188
Lombrozo, Tania, 95, 113–114, 120
Longino, Helen E., 8–9, 17, 70, 75–77, 110, 203, 205, 216, 229
Lotka-Volterra equations, 176

MacArthur, Robert H., 168
Machamer, Peter, 9, 25, 135, 164, 182
Mackie, J. L., 230

Maher, Brendan, 74, 75
Manuck, Stephen P., 71, 72
mate selection. *See* sexual selection
mathematical models, 70
Maull, Nancy, 194
May, R. M., 168
Maynard Smith, John, 66, 150, 169, 230, 233
McGill, Brian, 9, 162, 180, 183–184, 221
McGivern, Patrick, 9, 168
McKaughan, Daniel J., 205
McMullin, Ernan, 44
McShea, Daniel W., 232
mechanisms, 33, 135, 164–165, 169, 175, 176, 178–180, 182–183, 194
Melnyk, Andrew, 210
Merilaita, Sami, 172
metaphysics, 2, 16, 21, 161, 162, 165, 169, 174, 176, 186, 198, 212, 213
    and causation, 23, 32–34, 178
    and science, 21, 207–210
microphysics, 1, 82, 84, 166, 174, 180, 186, 187, 212
Milam, Erika, 233
Millstein, Roberta L., 9, 182, 233
mind-body relationship, 163
missing heritability, 11, 74, 76, 77, 216
Mitchell, Sandra D., 2, 16, 40, 164, 165, 172, 193–195
Mitchell, William A., 201
model-based science, 6, 18, 65, 84, 90, 93, 99, 193
model-target relationship, 52–54
molecular biology, 25, 74, 182, 187
molecular genetics, 71–76, 153, 192, 216. *See also* genetics
Molles, Manuel C., 166
Morgan, Mary S., 103
Morrison, Margaret, 103
multiple models, 85, 86, 92
mutualism. *See* cooperation

Nagel, Ernest, 165, 173, 188
natural kinds, 209
natural selection, 4, 10, 19, 36, 49, 50, 56, 60, 63–70, 78, 87, 88, 90, 99–101, 107, 111, 152, 155, 169, 176, 182, 184, 189–193, 200, 202, 212, 232
    group, 64, 65, 69
    individual, 65, 69
    kin, 64, 66, 69
    levels of selection, 10, 69, 169
Neurath, Otto, 188–189, 194, 197, 211
neurobiology, 72–75
neuroimaging, 72
Newton, Isaac, 24–26, 43, 50, 51
Ney, Alyssa, 33, 35

Obesity System Atlas, 37–40
objectivity, 8, 208, 210, 229
Odenbaugh, Jay, 92

Okasha, Samir, 10
O'Neill, Robert V., 168-169
Oppenheim, Paul, 1, 165, 166, 173, 174, 186, 188. *See also* philosophy of science, approaches to: completed science
optimality models, 31, 137, 142, 200-202
Oreskes, Naomi, 5
Orians, Gordon H., 166
Orzack, Steven Hecht, 10, 85
Owens, I. P. F., 190
Oyama, Susan, 15

Parker, Wendy S., 85, 87, 105
Pavlov, Konstantin A., 71, 73
Paynter, Nina P., 11
peacock train, 189-191
Pettit, Philip, 122, 141
phenotypic plasticity, 150, 151, 201
philosophy of science
  normative-descriptive dichotomy, 7-9
philosophy of science, approaches to
  completed science, 1
  feminist, 6, 8, 204
  general vs. field-specific, 8, 10, 19
  rational reconstruction of science, 1
  scientific practice, 1, 6-11, 18, 197, 219
  socially engaged, 6
physical chemistry, 187
physiology, 187
Pincock, Christopher, 143
Plaisance, Kathryn S., 6
pluralism, 16, 21, 199, 213-217, 221
Polger, Thomas W., 179, 230, 232
policy-making and science, 10, 86, 88, 105-111, 133, 204, 208, 217
population biology, 31, 49, 50, 56, 69, 110, 187, 189, 201, 202
population genetics, 10-11, 16, 53, 107, 194. *See also* genetics
pragmatics, 126
predator-prey relationships, 155, 176
prediction, 10, 19, 68, 70, 84-88, 90, 92, 105, 107-111, 117, 118, 189, 204, 205, 208, 217
prisoner's dilemma. *See* evolutionary game theory
progress in science, 58, 121, 198, 199, 218-219
Purves, William K., 166
Putnam, Hilary, 1, 122, 141, 165, 166, 173, 174, 180, 186, 188. *See also* philosophy of science, approaches to: completed science
pyramid view of science, 165, 168, 186

quantum mechanics, 19, 62, 82-84, 90, 100, 109, 121, 210, 211

rational economic agent, assumption of, 2, 43, 52
realism vs. antirealism, 20, 119

realization and multiple realization, 161, 164, 165, 169, 171-174, 179-180, 186, 195, 208-210
redshank sandpiper, 31, 32, 137
reductionism, 9, 10, 141, 161, 163, 165-166, 168-170, 173, 174, 186, 188, 195, 197, 209-210, 221
Rehnder, Bob, 95, 114, 120
relativity (physics), 210
relevance: causal and explanatory, 145
representation, 17, 19, 23, 25, 26, 28, 33, 36, 39, 40, 44-46, 48, 52, **52-54**, 55, 57, 60, 62, 67, 68, 90, 91, 93, 101, 105, 120, 123, 131-133, 162, 172, 180, 186 (*see also* idealization: representing as-if)
  and accuracy, 24, 25, 49, 51, 53, 58, 59, 66-70, 74, 75, 79, 85, 86, 90-93, 95, 97, 106, 107, 109, 110, 117-119, 202, 204, 215, 218 (*see also* explanation: and accuracy)
representation as-if, 144, 156
research programs, 150-153
Rice, Collin, 31, 44-47, 59, 92, 132, 138
Richardson, Robert C., 2
Ricklefs, Robert E., 166
robustness, 44, 85-87, 92, 106
Rohwer, S., 150
Rohwer, Yasha, 44-47, 59, 92, 132, 230
Rosenberg, Alexander, 80, 192
Ross, Don, 166, 210-212
Roughgarden, Joan, 4, 6, 65, 151, 152, 190, 193, 203, 220, 233
Rueger, Alexander, 9, 168

Sadava, David, 166
Salmon, Wesley, 31, 33, 129, 135, 231
San Francisco Bay Area Rapid Transit system, 51
Sazima, Ivan, 172
scale relationships, 221
  spatial, 162, 177-179, 183, 185, 186
  temporal, 177-179, 183, 185
Schoener, Thomas W., 168
scientific community, 8, 11, 16, 46
scientific knowledge, 208
scope of causal dependence, 39, 40, 136, 139-141, 144, 145, 148, 151, 152, 158, 160
sex differences
  aggression and promiscuity, 2-6, 17, 203
  Bateman's principle, 3-4
  genetics and IQ (*see* genetics: and IQ differences)
  mate selection, 3
sexual selection, 2-4, 6, 17, 189-191, 203, 220
Shapiro, Lawrence A., 179, 232
Short, R. V., 190
similarity. *See* model-target relationship
simulations. *See* computational models
Skipper, Robert A., 37, 182, 230
Snell's law, 96, 98, 106

Sober, Elliott, 10, 31, 69, 85, 138, 141, 231
socially engaged science, 221
social norms and science. *See* values and science
species and phylogenetics, 10
Sterelny, Kim, 9
Stevens, Martin, 172
Strevens, Michael, 9, 16, 17, 18, 58, 94, 114, 123–125, 129–130, 134, 135, 137, 140–142, 147, 155, 176
structural cause, **32**, 32, 36, 49, 136, 138, 142, 145
Suárez, Mauricio, 230
supervenience, 164, 165, 169, 173
Suppes, Patrick, 9
Szathmáry, Eörs, 169

Tabery, James, 77
Takahashi, Mariko, 189–191
tractability and modeling, 44, 47–49, 139, 216
Trivers, Robert L., 230
Trout, J. D., 104, 115, 231
true enough. *See* epistemic acceptability
truth, 20–21, 85, 89, 91, 93–96, 98, 101–103, 105, 109, 112, 113, 115, **117–120**, 120, 133, 192, 198, 199, 204, 206–208, 212–215, 218–219, 221

understanding, 20, 29, 34, 91, 101, 103, 104, **93–104**, 105, 107, 115, 118, 122, 127, 130–134, 153, 160, 186, 193, 198–201, 203, 204, 206–208, 210, 211, 213, 214, 217, 218, 221
  and belief, 96–97, 105
  and biases, 115
  dual nature: cognitive and epistemic, 94, 112–116, 131
  and generalizations, 113–114
  and objectivity, 104
  tacit, 132
unity of science, 162, 163, 170, 186, 188, 209, 212, 214–217
  coordinate unity, 188, 190–197, 221

Vaihinger, Hans, 230
Valone, Thomas J., 201
values and science, 2, 4–6, 21, 198–206, 213, 220–221
vampire bats, 215
van Cleve, Jeremy, 65
van der Waals equation, 26, 28, 29, 42, 98
van Fraassen, Bas C., 53, 126, 134, 149, 231

Waide, J. B., 168
Waters, C. Kenneth, 9
Weisberg, Michael, 44–47, 53, 58, 59, 85, 230
West, Geoffrey B., 168
why-questions, 126, 128, 149
Wilkins, Jon F., 194
Wilkinson, Gerald S., 63, 66–68
Williams, Joseph J., 95, 114, 120
Willmes, David, 96, 105, 199
Wilson, David Sloan, 69
Wilson, Edward O., 168
Wilson, Robert A., 171
Wimsatt, William C., 2, 6, 43, 92, 93, 116, 165, 230, 232
Winsberg, Eric B., 56, 85, 87, 96, 205
Woodward, James, 29–32, 34, 39, 40, 95, 134–136, 141, 143, 230
Worden, Lee, 65
Wright, Cory, 127

Zagzebski, Linda, 103

www.ingramcontent.com/pod-product-compliance
Lightning Source LLC
Chambersburg PA
CBHW021940290426
44108CB00012B/911